BURLEIGH DODDS SCIENCE: INSTANT INSIGHTS

NUMBER 18

Improving piglet welfare

burleigh dodds
SCIENCE PUBLISHING

Published by Burleigh Dodds Science Publishing Limited
82 High Street, Sawston, Cambridge CB22 3HJ, UK
www.bdspublishing.com

Burleigh Dodds Science Publishing, 1518 Walnut Street, Suite 900, Philadelphia, PA 19102-3406, USA

First published 2022 by Burleigh Dodds Science Publishing Limited
© Burleigh Dodds Science Publishing, 2022. All rights reserved.

British Library Cataloguing in Publication Data
A catalogue record for this book is available from the British Library

ISBN 978-1-80146-055-2 (Print)
ISBN 978-1-80146-056-9 (ePub)

DOI: 10.19103/9781801460569

Typeset by Deanta Global Publishing Services, Dublin, Ireland

Contents

4   Optimising the health of weaned piglets   121
Andrea Luppi, Istituto Zooprofilattico Sperimentale della
Lombardia e dell'Emilia Romagna (IZSLER), Italy

Series list

Title	Series number
Sweetpotato	01
Fusarium in cereals	02
Vertical farming in horticulture	03
Nutraceuticals in fruit and vegetables	04
Climate change, insect pests and invasive species	05
Metabolic disorders in dairy cattle	06
Mastitis in dairy cattle	07
Heat stress in dairy cattle	08
African swine fever	09
Pesticide residues in agriculture	10
Fruit losses and waste	11
Improving crop nutrient use efficiency	12
Antibiotics in poultry production	13
Bone health in poultry	14
Feather-pecking in poultry	15
Environmental impact of livestock production	16
Pre- and probiotics in pig nutrition	17
Improving piglet welfare	18
Crop biofortification	19
Crop rotations	20
Cover crops	21
Plant growth-promoting rhizobacteria	22
Arbuscular mycorrhizal fungi	23
Nematode pests in agriculture	24
Drought-resistant crops	25
Advances in crop disease detection and decision support systems	26
Mycotoxin detection and control	27
Mite pests in agriculture	28
Supporting cereal production in sub-Saharan Africa	29
Lameness in dairy cattle	30
Infertility/reproductive disorders in dairy cattle	31
Antibiotics in pig production	32
Integrated crop–livestock systems	33
Genetic modification of crops	34

Chapter 1

Optimising sow and piglet welfare during farrowing and lactation

Emma M. Baxter, Animal Behaviour and Welfare Team, Animal and Veterinary Sciences Research Group, SRUC, UK; and Sandra Edwards, Newcastle University, UK

1 Introduction

One of the most contentious and persistent welfare issues in livestock production is the use of housing systems involving confinement. In the pig industry, these include gestation stalls and farrowing crates that together house sows during much of their reproductive life. The welfare detriments to the sow include physical and behavioural restriction leading to physiological and psychological stress. There is growing pressure, expressed through consumer demand, societal opinion and government legislation, to abolish confinement systems such as the farrowing crate (Eurobarometer, 2016). However, farmers have valid concerns regarding such change; they need to achieve good animal performance (i.e. high piglet survival), in systems with acceptable capital, running and labour costs, and which facilitate efficient labour routines and safeguard the operator. Originally, farrowing crates were introduced to achieve these aims, especially to reduce piglet mortality attributed to overlaying by the sow. While reductions in this type of mortality were achieved, piglets also die from other causes (Edwards, 2002) and their mortality is a persistent welfare and economic problem (Baxter and Edwards, 2018). Narrow production-focussed breeding goals exacerbating many of the pre-disposing risk factors for piglet mortality have hindered improvements in survival and further challenge both piglet and sow welfare during farrowing and lactation.

http://dx.doi.org/10.19103/AS.2020.0081.04

There are also welfare concerns for piglets relating to painful husbandry procedures such as tooth resection, tail docking and castration. Furthermore, certain management strategies to rear surplus piglets from large litters pose new risks for piglet welfare. This chapter will highlight some of these welfare challenges, before describing various mitigation strategies to optimise welfare of both sows and piglets in the farrowing environment. Various examples of best practice will be detailed, with a case study describing the development and successful implementation of a high-welfare alternative farrowing and lactation system ('PigSAFE') that attempts to reconcile the 'triangle of needs' relating to the farmer, the sow and her litter.

2 Welfare challenges during farrowing and lactation

The most prominent welfare concerns during farrowing and lactation focus on systems imposing both physical and behavioural restriction on the sow. This is an area of continued scientific and public debate. 'Naturalness' and 'freedom to express natural behaviour' are facets of animal welfare that citizens value highly (Verbeke, 2009; Sørensen and Fraser, 2010; Boogaard et al., 2011) and surveys repeatedly emphasise that sow confinement is in direct opposition to these values (Boogaard et al., 2011; Grunert et al., 2018), with respondents wanting 'spacious farrowing pens with unfixed sows' (Boogaard et al., 2011). Despite cultural differences in attitudes towards animals (D'Silva and Turner, 2012), dissatisfaction with restrictive systems appears to be globally shared. European concerns are well documented (Eurobarometer, 2016) and have resulted in total or partial prohibition of some confinement systems such as sow gestation stalls. Studies in the USA and Canada (Ryan et al., 2015) have shown that North American consumers also oppose the use of stalls and therefore it is likely that they would raise similar objections to 'fixation' of the sow at any period in her life. Brazilian citizens preferred cage-free or free-range systems based on naturalness, animals' freedom to move, and ethics (Yunes et al., 2017). Although animal welfare is a relatively new concept in China, there are indications that the consumers here are also becoming increasingly interested in improving the rearing conditions of farm livestock (You et al., 2014; van de Weerd and Ison, 2019), especially if this impacts upon food safety. Younger consumers, in particular, value environmentally friendly and organic products (de Barcellos et al., 2013; Thøgersen and Zhou, 2012; Chen and Lobo, 2012), suggesting increasing pressure on confinement systems in the future.

While progress has been made worldwide in restricting the use of stalls in gestation (Buller et al., 2018), the predominant maternity system used globally continues to be the farrowing crate. It persists because its principal reason for development persists - high levels of pre-weaning piglet mortality, along with the perception that the majority of these deaths are attributable to overlying by the sow (Edwards, 2002). The farrowing crate reduces crushing mortality by significantly limiting the sow's freedom of movement and provides a safe and cost-effective working environment for farm staff. However, piglets can still die from other causes and the farrowing crate may not be the only solution. Average piglet pre-weaning mortality levels in UK commercial outdoor systems, where the sow experiences zero-confinement throughout farrowing and lactation, are on a par with those in indoor crated systems (12.20% vs. 12.19% respectively), with stillborn mortality actually lower in outdoor units (3.3% vs. 5.8%, AHDB Pork[1]). Furthermore, there is a growing body of evidence demonstrating that the reduction in sow welfare associated with confinement in a farrowing crate actually results in impaired welfare outcomes for her piglets. Confined sows show increased piglet-directed aggression (Jarvis et al., 2006), and greater restlessness during farrowing which increases crushing risk and reduces safe access to the udder (Ocepek and Andersen, 2017). Conversely, in loose-housing environments where sow welfare is improved, piglet advantages include increased maternal carefulness (Ocepek and Andersen, 2017), improved suckling success, as measured by increased IgG levels (Yun et al., 2014), and increased weaning weight (Pedersen et al., 2011; Nowland et al., 2019).

2.1 Piglet welfare challenges

The main welfare challenges for piglets are the pre-disposing factors for mortality. Piglet mortality during farrowing and lactation averages 16-20% per litter (Baxter and Edwards, 2018). Their most vulnerable period is during the first 72h of life, when they can suffer asphyxia, hypothermia, crushing and starvation. These early causes of live-born mortality are often interlinked, and many piglets will become chilled, fail to compete at the udder for vital colostrum and lack energy to move away from the sow when she changes posture (Edwards, 2002). Piglets can also suffer from disease pre-weaning, especially if they fail to suckle colostrum from which they gain passive immunity. Thus, many piglets that die pre-weaning are likely to be exposed to some degree of either pain, hunger and/or fear and stress which could be either acute or chronic. Piglets that are born dead are unlikely to have reached a conscious state and therefore stillbirth is less of a welfare issue for the piglet itself (for a more detailed discussion see Baxter and Edwards, 2018). However, with piglet mortality of approximately 7% attributed to this condition (annual benchmarking figures from commercial herds in USA, UK and DK for 2018, sourced from PigCHAMP, AHDB Pork, SEGES), it is certainly an economic and ethical concern. Many of

---

1 https://porktools.ahdb.org.uk/prices-stats/costings-herd-performance/ data acquired January 2020.

the pre-disposing risk factors for mortality have been exacerbated by narrow breeding goals focussing on production traits such as hyperprolificacy and lean tissue growth rate, which reduce piglet maturity at birth and increase within-litter competition both pre- and post-natally. A particular concern for piglet welfare is the increased number of piglets born with intrauterine growth retardation (IUGR) (Edwards et al., 2019a). This not only increases neonatal mortality risk, but also has detrimental long-term consequences for physiology and behaviour.

If piglets do survive early life, they will commonly undergo a number of painful husbandry procedures including tooth resection, tail docking and (for males) castration. All of these procedures are performed within the first 7 days of life, usually within 24h post-partum. Piglets may also receive vaccinations and injections of iron, vitamins and antibiotics. While each of these procedures is carried out with some longer-term welfare justification, all cause acute pain and can give rise to additional medium- and long-term welfare detriments.

Piglets can experience further behavioural detriments as a result of barren housing environments, with no access to environmental enrichment or structurally complex surroundings that are known to have positive effects on social and cognitive development (De Jonge et al., 1996; Martin et al., 2015), adaptation to weaning (Oostindjer et al., 2011a), growth rate (Brown et al., 2015; Lawrence et al., 2018), immune responses (van Dixhoorn et al., 2016) and stress regulation mechanisms (Fox et al., 2006). Piglets reared in barren environments often develop poor social skills, display abnormal behaviours both pre- and post-weaning, and lack behavioural flexibility to cope with challenges later in life (for a review see Telkänranta and Edwards, 2018).

2.2 Sow welfare challenges

In conventional pig production, sows are moved to farrowing crates approximately five days before they are due to give birth. A typical farrowing crate has tubular metal bars running horizontally along its length, with additional bars positioned above the sow to prevent escape by jumping or climbing. It measures approximately 2.00m in length and is between 0.45-0.65m wide (Table 4). Flooring is typically fully or partially slatted (plastic or metal) to allow easy removal of waste into slurry pits below. The farrowing crate restricts movements, allowing the sow only enough space to stand up and lie down but not to turn around. The flooring and manure management system generally prohibit the provision of substrate required to help fulfil behavioural needs, such as the performance of highly motivated nest-building behaviour (Wischner et al., 2009). Nest-building is a behavioural pattern typically initiated by sows from 16-24h before they give birth. Its performance reflects a strong evolutionary survival value since, in the wild, these nests protect the young from predators and inclement weather. Nest-building is therefore intrinsically motivated; triggered by endogenous hormones (Algers and Uvnas-Moberg, 2007) and stimulated by extrinsic environmental factors in a sequence including finding a nest-site, rooting the ground, finding, carrying and arranging suitable substrate (Jensen, 1986, 1993). The time sequence in which nest-building behaviour ceases and farrowing starts is influenced by the performance of these activities and the effects that they have on hormone levels that prepare the sow for farrowing. A satisfactory phase of active nest-building occurring 3-7h pre-farrowing coincides with increases in oxytocin level. The sow then becomes less active, lies down and goes into a 'quiet phase' before farrowing begins.

A sow will attempt to perform nest-building behaviour no matter what her environment. Sows kept in farrowing crates or pens with no nest-building material redirect these behaviours to the pen equipment and perform bar-biting, manipulate the drinker and root and paw at the floor (Lawrence et al., 1994). A common misconception is that modern domestic breeds of sow do not nest-build as they have no need, experiencing no risk from predators or inclement weather when housed indoors, and because the behaviour has been bred out of them. This is incorrect and nest-building remains functionally important. Despite years of domestication and selective breeding, and despite protective, warm environments provided to piglets, the modern sow continues to be highly motivated to perform the behavioural patterns of nest-building (Jensen, 2002; Yun and Valros, 2015). When sows are unable to fully perform nest-building due to a lack of space and/or substrate, they show increased stress (Lawrence et al., 1994, 1997; Jarvis et al., 1997, 2001). This adversely affects various aspects of maternal behaviour (for a review see Yun and Valros, 2015) including the progress of farrowing, which has been shown in some (Oliviero et al., 2010), but not all (Jarvis et al., 2001; Hales et al., 2015), studies to increase stillborn mortality. Stress can activate brain opioid pathways, resulting in an inhibition of oxytocin release which is essential to the farrowing process (Lawrence et al., 1992). Opioids play an important role in preventing pain during farrowing (Jarvis et al., 1998; Ison et al., 2018) and may also promote sow passivity towards piglets during the early post-natal period (Jarvis et al., 1999). Disruption of the hormonal control of this passive response may trigger piglet-directed aggression (Ahlström et al., 2002), which is more prevalent when sows are housed in farrowing crates (Jarvis et al., 2004; Ison et al., 2015). Furthermore, reductions in positive maternal behaviours including sustained lateral lying, carefulness when changing posture, including pre-lying sow-piglet communication and responsiveness to piglets are also reported in restrictive systems (Jarvis et al., 1999; Andersen et al., 2014; Yun and Valros, 2015).

Other physiological and physical impacts of confinement on the sow include a reduced ability to thermoregulate (Quiniou and Noblet, 1999; Phillips et al., 2000; Muns et al., 2016), increased risk of hoof, leg and shoulder lesions (Boyle et al., 2002; Leeb et al., 2001) and reduced muscle mass due to prolonged reduction in movement (Barnett et al., 2001). Restrictive housing and the impairment of thermoregulation leads to reduction in feed intake (Pajor, 1998; Muns et al., 2016; Black et al., 1993), which contributes to the development of shoulder lesions/ulcers (Bonde, 2008) and adversely affects milk production and piglet growth rate (Black et al., 1993; Renaudeau and Noblet, 2001). The modern hyperprolific sow is at greater risk of developing injuries when housed in farrowing crates due to her significant increase in size; she is now substantially heavier and longer than her equivalent of 40 years ago (Moustsen et al., 2011). If used as a nurse sow to rear surplus piglets or piglets weaned at an older age, an extended lactation length also results in increased incidence of leg bursae and udder lesions (Ladewig et al., 1984; Sørensen et al., 2016). Research shows that prolonged confinement in a farrowing crate leads to chronic stress in sows (Cronin et al., 1991; Jarvis et al., 2006). This appears to arise from enforced subjection to increasingly persistent piglet attention (Weary et al., 2002) and a lack of ability to express the exploratory and foraging behaviours seen under more extensive conditions as the sow increasingly spends more time away from the nest (Hötzel et al., 2004). Piglets housed in commercial farrowing crates spend more time interacting with their dam, including suckling and udder massage, than piglets reared outdoors (Hötzel et al., 2004). Therefore, it is likely that a lack of exploratory opportunities within conventional farrowing crates also contributes to welfare problems associated with this system.

3 Mitigating welfare challenges

Genetic selection programmes targeting piglet survival traits, including increased robustness and resilience of the neonate, as well as selection for good maternal traits (e.g. udder quality – Balzani et al., 2016; maternal behaviour – Ocepek and Andersen, 2017) would help to tackle some of the welfare detriments described previously. These strategies are discussed elsewhere (e.g. Turner et al., 2018). This chapter will concentrate on optimising nutritional inputs to help combat piglet mortality, as well as describing how changes to the farrowing and lactation environment can improve welfare for both the sow and piglets, while being mindful of practical considerations for the farmers.

3.1 Optimising nutritional inputs

3.1.1 Sow nutrition

There are basic nutritional requirements for sows and piglets that should be met to ensure, at least, provision of the nutrients required for maintenance and production outputs. However, research suggests that the nutritional needs of hyperprolific sows are not always met. Breeding for very large litter sizes imposes more demands on sows, and a need to update nutrient requirements (Strathe et al., 2017). For example, Danish hyperprolific sows have required increases in levels of lysine (by 28%) and crude protein (by 14%) in their rations (Tybirk et al., 2012; Tybirk et al., 2015 - cited in Pedersen et al., 2019). There is also evidence that modern hyperprolific sows have a compromised energy status at the time of farrowing (Feyera et al., 2018), in part due to a prolonged farrowing duration. Farrowing is now reported to last, on average, 7h (Hales et al., 2015), representing an increase of 150 minutes compared to their 'standard' counterparts producing 'normal' sized litters (Olivierio et al., 2010). Protracted farrowing can lead to dystocia, increased stillbirths and piglets born with hypoxia (Björkman et al., 2017; Olivierio et al., 2010), pain for the sow (Mainau et al., 2012; Ison et al., 2016), as well as reduced colostrum yield (Hasan et al., 2019). Given this extended farrowing period, it is likely that sows are suffering from maternal fatigue. Feyera et al. (2018) demonstrated that the time delay between a sow's last meal and the onset of farrowing affected farrowing duration and the number of stillborn piglets. If the time between the two events was below 3h there was no effect on farrowing duration or stillborn mortality, but if the delay was above 6h both were significantly affected. The risk of this occurring is high, given that sows prefer to farrow undisturbed overnight when staff are absent. Feyera et al. (2018) went on to show that the energy status of the sow was depleted as farrowing progressed and recommended that sows were fed three meals a day to reduce the inter-meal interval. They also demonstrated benefits from high-fibre diets, fed from two weeks prior to farrowing and over the transition period, which provided a more sustained energy release after feeding. Feeding high-fibre diets also provides sows with greater gut fill and feelings of satiety (D'Eath et al., 2009), and also relieves constipation (Olivierio et al., 2010), and has benefits for reproductive efficiency when fed post-weaning (for a review, see Jarrett and Ashworth, 2018). Providing sows with pain relief could also mitigate some of the welfare detriments associated with problematic farrowings. Mainau et al. (2012) demonstrated that sows treated with a non-steroidal anti-inflammatory drug (NSAID – meloxicam) spent less time lying inactive on the third day following parturition. They also demonstrated benefits for piglets (increased IgG concentration in serum and enhanced pre-weaning growth) when multiparous sows were given oral meloxicam at the beginning of farrowing (Mainau et al., 2016). Similarly, Homedes et al. (2014) saw a reduction in piglet mortality when sows were given the NSAID ketoprofen.

Other nutritional interventions at the sow level can indirectly influence piglet welfare. These include late gestational feeding of supplements to reduce the impacts of low birth weight on piglet vitality (e.g. palm oil distallate - Amdi et al., 2013a; fatty acids - Rooke et al., 2001; Bontempo and Jiang, 2015;

L-glutamine - Wu et al., 2011). The evidence for improvements in colostrum yield from dietary supplements is conflicting (Theil et al., 2014), but there appears to be greater success in altering colostrum composition via sow nutrition to provide more fat and energy to the piglet (e.g. supplementary leucine metabolite - Nissen et al., 1994; increased dietary fibre – Loisel et al., 2013). Introducing fats and oils into late gestation and lactation diets as a high-energy supplement has been shown to increase sow milk yield, improve neonatal growth and development, and alter sow metabolism (Bontempo and Jiang, 2015). Peng et al. (2019) showed that sows fed a gestation diet supplemented with fat (2% soybean oil) had greater plasma prolactin concentrations, which was thought to be the mechanism benefiting their lactation capabilities. The same study demonstrated increased colostral protein concentration, with others also showing soy oil supplementation in sow lactation diets increases protein concentration in milk (Jones et al., 2002).

3.1.2 Piglets: ensuring optimum colostrum intake

Piglets are completely reliant on their mother for the provision of nutrition in the first few days of life. In order to survive they must quickly get to the udder, acquire a functional teat and suckle colostrum. This not only aids thermoregulation and the acquisition of nutrients and immunoglobulins, but also initiates gut closure, which then reduces the risk of pathogens entering the piglet's systemic circulation (Gaskin and Kelley, 1995). With large litter size, the risk of failing to acquire enough colostrum for survival, growth and development is high (Quesnel et al., 2012; Decaluwé et al., 2014) due to competition at the udder and/or compromised birth weight status (Declerck et al., 2015; Hasan et al., 2019; Amdi et al., 2013b). Piglets need to ingest a minimum of 200g colostrum for survival, but at least 250g is required for normal growth and development (Devillers et al., 2011; Hasan et al., 2019). There is also evidence that, for successful transfer of lymphocytes (B and T cells), cytokines, nucleotides, various growth factors (Bandrick et al., 2011) and other important biochemical signalling factors (Power and Schulkin, 2013), piglets need to suckle colostrum from their own mother. This has implications for optimal farrowing management, and limits the success of early nutritional interventions providing colostrum supplements (e.g. Muns et al., 2014). It is likely that the energy piglets acquire from any supplements has more effect on piglet survival than the provision of immunoglobulins (Thorup et al., 2015; Muns et al., 2015, 2017). This is because the newborn piglet's body reserves do not cover its high-energy needs (e.g. to ensure thermoregulation, locate the udder, compete for a teat and escape potentially dangerous movements of the sow). Providing piglets with extra energy shortly after birth to boost activity could be a means to promote suckling (i.e. the acquisition of sufficient amounts of colostrum) and, consequently, neonatal survival and growth. However, it is difficult to suggest one protocol describing how best to deliver supplements, of which type and to whom, as studies vary in all these factors (e.g. Muns et al., 2014, 2015, 2017; Schmitt et al., 2019a; Englesmann et al., 2019). It is likely that, to be effective, energy supplements should be administered very early in the piglet's life and in not too great a volume in any one meal. This has implications for farrowing management and how best to design a farrowing environment that allows such targeted interventions while still safeguarding sow welfare. Certainly, maximising sow lactational output should be a priority and changes to the farrowing environment to influence colostrum quality (Yun et al., 2014) and provide easy access to the udder (Pedersen et al., 2011) have demonstrated positive effects on suckling success and piglet outcomes.

It has long been established that optimum birth weight is key to piglet survival (Tuchscherer et al., 2000; Edwards, 2002; Baxter et al., 2008) and the long-term prospects for piglets born under that optimum weight are poor because of their impaired suckling success. Roehe (1999) demonstrated that 1.6 kg was optimum, with piglets below this weight suffering a three-fold increase in mortality risk. Twenty years later, selection for hyperprolificacy has increased litter size and reduced average piglet birth weight, and it appears that there has been a shift in this optimum. A multi-study project, measuring birth weight of over 4000 piglets from 394 litters at four different commercial farms, concluded that piglets born below 1.11 kg suffered a six-fold increase in mortality risk (Feldpausch et al., 2019). The shift in average birth weight accompanying hyperprolificacy has not resulted in small but robust piglets more capable of survival. It is more likely that increased husbandry interventions for large litters have influenced the survival prospects of these smaller piglets. This is supported by recent evidence suggesting that, for piglets born under 1.8 kg in weight, it is their weaning weight that can best predict their life-time growth rate (Douglas et al., 2013), indicating that such piglets can display catch-up growth if provided with the correct resources. Therefore, ensuring high milk intake during the pre-weaning period for smaller piglets may help them to reach their longer-term growth potential. Supplementary milk provision in the farrowing pen, in addition to a stable suckling environment, may aid this endeavour. However, it should be noted that these interventions concentrate on growth and survival. Long-term health and welfare implications for piglets born into large litters and with low birth weight are yet to be fully understood (Edwards et al., 2019a).

3.1.3 Large litter management and ensuring optimum milk intake

If piglets are born into a very large litter and there are more piglets than functional teats, spilt suckling can be performed to ensure all piglets get as much colostrum as possible from their own mother (Baxter et al., 2013). This involves splitting the litter into two groups, usually based on their weight and/or vitality (Kyriazakis and Edwards, 1986; Donovan and Dritz, 2000). The lightest/weakest piglets are allowed access to the udder first, while the heaviest/strongest ones are enclosed in a heated creep area or a designated box. The piglet groups are then alternated after a few successful sucklings and this continues until fostering opportunities present themselves.

Cross-fostering can ensure a more even number and weight distribution of piglets on the udder to reduce competition and, if performed correctly in the first 48h of life (after a recommended 6h minimum with their own mother), it can be successful in promoting greater survival rates and growth of low birthweight piglets (Alexopoulos et al., 2018; Douglas et al., 2014). However, if performed too early (before 6h post-farrowing), too late (after 48h post-farrowing - Price et al., 1994; Straw et al., 1998; Horrell and Bennett, 1981) or over-performed it can be disruptive, stressful and counterproductive (Robert and Martineau, 2001; Straw et al., 1998). Piglets develop a teat order within the first few days of life, where they become faithful to an individual teat and will vigorously defend it (Puppe and Tuchscherer, 1999). Teat order stability comes from cohesive group suckling behaviour (Skok and Škorjanc, 2014) which promotes growth and development of piglets. Cross-fostering can disrupt this stability, and the introduction of new piglets into a litter that has already established its teat order can lead to fights causing facial injuries to piglets, sow udder injuries and disrupted milk ejections, which in turn can impact growth and development of both fostered and resident piglets (for a review, see Baxter et al., 2013).

Simple cross-fostering cannot solve the problem of supernumerary piglets in a whole farrowing batch, which is now commonplace in hyperprolific herds. More complex strategies need to be adopted to deal with these extra piglets throughout lactation. These include nurse sow strategies (i.e. use of sows that have had their own piglets early weaned in order to free up spare teats to rear surplus piglets) or artificial rearing of surplus piglets. Both practices are reviewed in detail in Baxter et al. (2013, 2018) and have implications for both piglet and sow welfare (Sørensen et al., 2016; Schmitt et al., 2019b). For nurse sow strategies, some studies (Kobek-Kjeldager et al., 2020) but not all (Schmitt et al., 2019c) show detrimental impacts on piglet weaning weight. Schmitt et al. (2019c) did find significant weight reduction in the first week after fostering when comparing piglets moved to nurse sows with piglets left with their mother, but by weaning the differences were no longer significant. They selected the heaviest and most robust piglets for fostering and chose nurse sows with good temperament and udder quality, which are likely to be factors in how successful these strategies can be. It should be noted that one major threat to piglet health is introduced by use of nurse sows; that is, the compromise of biosecurity by breaking batch integrity in the farrowing system (Calderón Díaz et al., 2017).

Minimising exposure of suckling piglets to pathogens is an integral part of controlling pre-weaning mortality and 'all-in-all-out' (AIAO) management of farrowing batches is the key to this. As nurse sow strategies involve some form of early weaning (at least relative to EU regulations), there are also implications for piglet welfare because of the known stressors involved in early separation from the dam (Drake et al., 2008).

It can be argued that artificial rearing is worse for piglet welfare than nurse sow strategies, since it typically involves very early separation from the mother and removal to a more barren environment. Here piglets are placed in specialised enclosures, usually located in a separate room or sitting above the farrowing crate, where they will be fed milk replacer until weaning age (usually at 28 days) (Schmitt et al., 2019d). The enclosures also contain a heat lamp to ensure thermal comfort of the piglets, milk and water cups that can be activated by nudging with their snout, and solid 'creep' food. The fact that piglets are fed ad libitum in a controlled environment, where the risk of crushing is removed, is quite attractive to farmers who may not be able to implement nurse sow strategies. Additionally, some studies have shown artificially reared piglets to have higher weaning weights (van Beirendonck et al., 2015; Cabrera et al., 2010). However, others have observed short-term (De Vos et al., 2014) or sustained (Schmitt et al., 2019d) impairments of growth in artificially-reared piglets as well as behavioural abnormality.

3.1.4 Pre-weaning solid feed intake

Piglets kept under natural or semi-natural conditions begin to forage for solid feed from about two weeks of age and gradually increase their intake as the sow's milk production declines. Introducing piglets to solid feed pre-weaning can reduce weaning stress, which can otherwise contribute to compromised gut and immune function, reduced feed-intake, growth rate, health and welfare (Weary et al., 2008). While early introduction of a starter diet is typically practiced on commercial farms from the second week of life, further optimisation could consider the palatability and presentation of the feed provided. In a series of experiments by Oostindjer et al. (reviewed in Oostindjer et al., 2014) it was demonstrated that post-weaning feed intake was improved when piglets could eat together with their mothers pre-weaning, and that flavour learning was also effective in promoting growth and reducing weaning stress. In these studies, the authors designed a shared feeding station allowing piglets to eat with their mother (Oostindjer et al., 2011b) and gave sows flavoured gestational feed before presenting piglets with the same flavour in both the feed and the air post-weaning (Oostindjer et al., 2011c). Middelkoop et al. (2018) also studied the impact of altering food presentation and flavour on piglet feed intake pre- and post-weaning. As early as two days post-farrowing, one group of piglets was presented with a choice of two diets differing in production method, size, flavour, ingredient composition and nutrient profile, smell, texture and colour (i.e. dietary diversity – DD). The other group had feed that was changed in flavour every six days (i.e. flavour novelty – FN). The DD piglets were more interested in feed and had higher intake than the FN group and it was thought that the diversity stimulated feed exploration and therefore intake. The same authors also demonstrated that the type of creep feeder influenced feed intake (Middelkoop et al., 2019). A round conventional feeder was less attractive to the piglets than a play-feeder, whereby the same round design was adorned with enrichment materials to stimulate rooting and exploratory behaviour.

Providing an enriched neonatal environment can also be beneficial for pre-weaning growth rate. Oostindjer et al. (2010) showed that piglets developed feeding habits better in a loose-housed farrowing and lactation environment compared to a conventional crate system. This is supported by evidence that an enriched neonatal environment that stimulates play behaviour can improve growth of litters (Brown et al., 2015). A combination of nutritional and environmental enrichment may be particularly valuable in mitigating the effects of being born with low birth weight. Understanding the importance of environmental enrichment is one aspect of a larger discussion about providing for the biological needs of the animals to optimise their welfare and enhance performance.

3.2 Biological specifications for optimal housing design

Although the definition of animal welfare has historically separated the animal's physical health and well-being from its psychological health (i.e. the 'biological functioning' vs. the 'feelings' schools of thought – Duncan, 2005), providing for both is imperative for good animal welfare. If a 'need' is denied, this results in a negative welfare state (Jensen and Toates, 1993). Where a need is life-sustaining (e.g. food and water), prolonged failure to make provision for it would be fatal. Therefore, hierarchies of need have been proposed (Duncan, 2005) with these basic life-sustaining needs understandably prioritised. Within livestock production, behavioural needs are often demoted or ignored within the hierarchy, perhaps because of a failure to recognise (or for animal welfare scientists to communicate) their importance for biological functioning. This may actually be counter-productive for animal productivity, as it is well established that performance of species-typical behaviour contributes to an animal's biological fitness (Hamilton, 1964a,b). Behaviours (e.g. nest-building) that have persisted despite centuries of domestication and selection pressure for production traits are those that remain biologically significant.

Thus, biological needs should include physiological, physical and behavioural necessities, and providing for these should promote welfare, health and performance. Meeting biological needs in the farrowing and lactation environment requires the identification of what is most important for both piglets and sows, and then determining the best way to make provision. When designing a new housing system, this means understanding that the animals themselves are active participants in the process and not just passive receivers of the husbandry systems they are put into, as typically practiced since the industrialisation of farming (Bos and Koerkamp, 2009). Many of the health, welfare and environmental problems facing livestock sectors today arise from the dominant post-war policies of modernisation, which stem, understandably, from the focus on ensuring domestic food security at an affordable price. Re-designing the environment with the occupants in mind should redress the balance.

In the case of farrowing and lactation, Baxter et al. (2011a) attempted this approach by first describing each physiological and behavioural need, as indicated by sound biological evidence, and then determining which of these needs are sensitive to the physical environment. They subsequently translated this information into design criteria for a maternity unit built to the biological specifications of the pigs (as summarised in Table 1). This initial task formed the basis for re-designing the farrowing and lactation environment to maximise welfare (The PigSAFE project – see Section 3.4).

3.2.1 Space

The sow needs space for finding and creating a nest, giving birth (parturition), suckling and interacting with her piglets, feeding, drinking, defaecation as well as space for re-integration with her social group. The piglet's spatial needs include space in which to be born, search for the udder, find a teat and suckle. It needs space to rest and keep warm, interact with its siblings and mother, as well as to avoid her during posture changes, and to grow and develop. With current genetics, this means providing space for at least 14 piglets (or the likely maximum number of piglets a sow could support) until weaning at 28 days post-partum.

Quantifying the spatial needs of the animals involves measuring the size of the modern-day domestic sow and her piglets (at weaning) based on the 95th percentile (i.e. largest) individuals. Then, determining both the static and dynamic space requirements for different postures and posture changes, turning around, piglet gathering, separation of functional areas, etc. (Baxter and Schwaller, 1983; Petherick, 1983, Gonyou et al., 2006). Allometry, based on the equation $A = k*W^{0.667}$ (where A=area, k= a constant and W= live-weight), can be used to estimate the space an animal occupies as a consequence of its mass (Baxter and Schwaller, 1983). This equation is very general and assumes all animals are the same shape and that this is consistent over time. In general,

Table 1 Summary of sows' and piglets' biological specifications for a farrowing and lactation system and estimated 'values' required to meet their needs. More detailed descriptions are given in Baxter et al. (2011a, 2018) and any new information published since these manuscripts has been incorporated. Where no new information has materialised and values for components cannot be given, 'further research' is noted

Component of system	Sows	Value of required specification at minimum level to meet needs	Piglets	Value of required specification at minimum to ideal level
Space	Increased activity for nest-site seeking	4.9m²	Parturition	2.79m²
	Hygiene – separate dunging space from feeding and lying area	Separate dunging area from nest and feed sites.	Udder access for suckling throughout lactation	2.79m²
	Feeding and foraging	Separate feeding area from nest and dung sites	Protection, safe lying area for parturition and nest-occupation	Separate space inaccessible to the sow e.g. 0.8m² per 10-12 neonates
	Turn-around nest space for piglet inspection and gathering behaviour	Floor space = 2.44m², planar space = 3.17m². Further research needed		
	Lateral lying and parturition	2.79m²	Protected lying area during lactation	0.96m² for 14, four-week-old piglets
	Thermal comfort via posture changes	2.44m²	Area for feed trough to introduce starter diet and area for supplying supplementary nutrition/energy (separate from the sow)	Provide in the creep, interacts with above
	Nest-departure	Separate area from nest site		
	Gradual separation from piglets and sow-controlled nursing	Separate space inaccessible to piglets	Hygiene	Separate area for dunging, interacts with flooring
	Social contact with other sows	Separate space inaccessible to piglets to allow contact between sows, but if sows fully reintegrated before weaning larger space to allow body language assessment and/or fighting to establish dominance – further research needed to determine minimum space per sow.		

Category	Feature	Function	Recommendation	Notes
Substrate	Nest-building – carrying and manipulating	Foraging, nutritional development	2 kg long-stemmed straw	Earth-like materials (e.g. peat). Further research needed on quantity. Novelty requires fresh input daily. Complex materials (e.g. branches) preferred.
	Complete nest	Enrichment, social and cognitive development	2 kg long-stemmed straw and branches	2.5cm of straw, interacts with flooring
	Udder comfort	Thermal comfort during parturition	Further research needed, interaction with floor properties	Further research needed, interacts with thermal comfort and flooring properties
	Thermal comfort during nest-building and parturition	Physical comfort	2 kg long-stemmed straw in nest site – interaction with flooring and room temperature.	Deep bedding – 10-12cm, interacts with flooring
	Foraging material	Protection	Further research needed	
Walls	Enclosure/Isolation of nest	Protection from sow posture changes	3 solid-sided walls (cul-de-sac)	Sloped wall or protection bars
	Visual and physical contact with non-litter pigs	Social contact (visual and physical)	Vertical barred area with void wide enough to allow at least nasal contact between pigs	Vertical barred area
		Hygiene	Solid at base with separation between pens	Solid walls (at least at bottom of penning) separating other litters
	Supported posture changes	Thermal comfort	Solid sloped or vertical walls	Solid walls with thermal resistance properties to limit heat loss via radiation – interacts with substrate and flooring
	Lack of disturbance		Further research needed	

(Continued)

Table 1 (*Continued*)

Component of system	Sows	Value of required specification at minimum level to meet needs	Piglets	Value of required specification at minimum to ideal level
Flooring	Nest-building - digging, rooting and hollowing	Malleable (e.g. earthen) or solid to accommodate deep substrate	Thermal comfort during parturition and first 24h of life	High thermal resistance - e.g. rubber matting or deep substrate or under-floor/localised heating
	Nest-building and parturition	Solid to accommodate substrate	Thermal comfort during lactation	High thermal resistance - e.g. rubber matting or deep substrate (see above) or under-floor/ localised heating (see general)
	Thermal comfort during nest-building, parturition and lactation	Temperature differentials in separate areas allowing choice.		
		High thermal resistance e.g. rubber matting or deep substrate. Low thermal resistance e.g. metal.	Physical comfort - avoiding injury, promoting suckling behaviour	Solid flooring with minimal abrasiveness and well-maintained (e.g. rubber matting or specialised screed with non-slip properties), interacts with substrate
			Protection from fatal crushing by the sow	
	Physical comfort - avoiding injury, promoting suckling behaviour	Non-slip surface e.g. rubber matting or plastic-coated metal	Hygiene	Slatted flooring with void width no more than 10mm and rounded edges. Interacts with temperature (see general)
		Minimal abrasiveness (interacts with substrate). Solid to avoid teat injuries		
	Hygiene	Slatted area		
		Gradation of floor with slope away from lying area. E.g. 10% slope for drainage		

General		
Thermal comfort	Ambient temperature 12–22°C, interactions with substrate and flooring	
High feed intake	See space and thermal comfort	
Low light in nest-site		
	Health treatment for injuries, vaccines, etc.	Safe area for handling required, interacts with space
	Promote weaning, reduce nutritional stress and encourage increased feed and water intake	Suitable solid food, inaccessible to the sow – interacts with space and substrate. Provide feed tray and sufficient space to allow social facilitation
	Thermal comfort	Localised heat source set at thermo-neutral temperature (e.g. 34°C at birth) – interacts with substrate
	Hygiene	Temperature differentials to encourage dunging outside of nest site – interacts with flooring

because these equations err on the side of caution, they are still relevant today for calculating static space requirements (Gonyou et al., 2006). However, many systems are so fixed in their pen dimensions that they have provided no flexibility for coping with changes in the animal's size and productivity over time, or changes to husbandry practices that impact on space. The farrowing crate is such a system. Because the modern-day sow is substantially larger than her counterparts of the 1980s (Moustsen et al., 2011) and she rears a larger litter for a longer lactation length, many farrowing pen designs do not allow sufficient space for hyperprolific litters (Pedersen et al., 2013). Piglets can neither rest together in a thermally comfortable area nor have unobstructed suckling. EU Council Directive 2008/120/EC states, 'When a farrowing crate is used, the piglets must have sufficient space to be able to suckle without difficulties'. The length of a four-week-old piglet (approximately 0.5 m – Moustsen and Poulsen, 2004), therefore needs to be provided at each side of the crate to avoid fighting caused by blocking of piglets' access to preferred teats. Suckling success is promoted when such constraints are not in place, with greater milk let-down reported when sows show better udder access in loose housing (Pedersen et al., 2011). The detailed body measurements provided by Moustsen and colleagues allow calculations for the necessary creep dimensions, suckling space and space for fixtures and fittings within a pen to prevent injury (e.g. creep bar widths). These data can then be applied to determine the detailed design dimensions for any new farrowing and lactation system (Table 1).

3.2.2 Substrate

Although there is strong evidence that space is more important than substrate for allowing the behavioural expression of nest-building (Jarvis et al., 2002; Cronin et al., 1994; Hartsock and Barczewski, 1997), substrate is still very important for sows and piglets. Given that the majority of commercial sows are housed in conventional crated environments (>90% in selected European countries, >85% in Australia and New Zealand – Baxter and Edwards, 2016[2]), it is important to investigate more fully which substrates are of relevance to the sow before and after farrowing in order to improve her welfare, maternal behaviour and the welfare of her piglets. In the periparturient period, the sow needs to collect and arrange suitable substrate for nest-building, and to achieve thermal and physical comfort for herself and her piglets. Substrate also facilitates foraging and nutritional development, providing enrichment which is important for the social and cognitive development of piglets.

---

2 Proceedings of Free Farrowing Workshop 2016, Belfast NI (eds, Baxter, E. M. and Edwards, S. A.). Proceedings available at https://www.freefarrowing.org/info/2/research/45/free_farrowing_workshops.

The type of substrate provided should have properties which satisfy the biological needs. For nest-building, the material needs to be suitable to provide feedback to the sow signalling that the nest is complete. Without this, some sows may continue to be motivated to nest-build even during farrowing (Thodberg et al., 1999; Damm et al., 2000), which constitutes a risk for the newborn piglets. Substrate manipulation to allow arrangement, as well as a sense of enclosure, appears to give a sense of completion and, in preference tests, sows chose pre-formed nest-sites that also offered a sufficient quantity of straw to satisfy nest-building behaviour (Arey, 1992). 'Sufficient quantity' is a rather open description but it is one found in the European legislation outlining provision of enrichment material at all stages of production. Council Directive 2008/120/EC states that 'pigs must have permanent access to a sufficient quantity of material to enable proper investigation and manipulation activities'. For nest-building, studies have suggested as little as 2 kg of long-stemmed straw to be sufficient (on solid floors – L. J. Pedersen *personal communication*), whereas voluntary use of as much as 255 kg of mixed substrate has been reported (outdoor semi-natural systems - Zanella and Zanella, 1993). Practicality is important and interactions with the flooring and waste management system are factors requiring consideration. EU legislation says that all sows must have material for nest-building but provide possibilities for derogation, stating that 'unless the slurry system makes provision unfeasible'. Straw is widely considered as the gold-standard material for its manipulable as well as thermal properties (Rosvold et al., 2018; Mount, 1967; Wathes and Whittemore, 2006). Westin et al. (2014) demonstrated that, if used strategically, a slurry system could cope with large quantities of straw at the time of nest-building. They provided 15–20 kg of long-stemmed straw when sows moved in to the farrowing accommodation and did not subsequently remove (unless soiled) or add to this, allowing the straw to gradually fall through the slatted part of the pen and be removed with the excreta. Alternative nesting materials have also been studied (e.g. cloth tassels - Widowski and Curtis, 1990; burlap sacks – Bolhuis et al., 2018; Plush et al., 2019; lucerne hay - Edwards et al., 2019b; peat - Rosvold and Andersen, 2019; newspaper, sawdust, shredded paper - Swan et al., 2018). Something as simple as a cloth tassel tied to pen fittings that animals can pull, tear and manipulate may provide welfare benefits, even though the material cannot result in the building of a suitable nest. This is particularly relevant for animals kept in crates, as most of the tassel remains attached to the front of the crate where sows can continue to access it. In contrast, substrate on the floor often gets pushed to the rear of the crate or out of reach during nest-building, with potentially frustrating consequences. Swan et al. (2018) specifically looked at nest-building in crates and compared six materials, either attached to the farrowing crate or in amounts of 1–2 L placed in front of the sow. These included point-source objects offered on the ground or to the side, wood shavings, straw, shredded paper and whole newspaper. Sows showed nest-building behaviour with all materials, but a functionality assessment saw straw, wood shavings and newspaper tested further. There were different benefits for the selected options: the newspaper group performed more nest-building and fewer bar-biting activities, while piglet mortality during the entire lactation period was lower in the straw group than the other groups. This study demonstrates benefits to the piglets of maternal nest-building and is supported by studies, both in crates and loose-housing, that show providing for nest-building can improve suckling success and growth rate in piglets (e.g. crates - Edwards et al., 2019b; loose housing - Yun et al., 2014; Pedersen et al., 2011; Plush et al., 2019) and can reduce stillbirths (e.g. Rosvold and Andersen, 2019; Edwards et al., 2019b). Edwards et al. (2019b) also showed that sows in crates continued to interact with provided substrate (in the form of lucerne hay) throughout lactation, suggesting they still found the enrichment rewarding after the nest-building phase.

Enriching the environment throughout lactation will also benefit the piglets. Studies have shown that providing substrates, point-source objects, opportunities for play and socialisation improves cognitive and social skills (Martin et al., 2015), growth rate (Brown et al., 2015), gut health (Oostindjer et al., 2010) and immunity (van Dixhoorn et al., 2016), and can reduce weaning stress. Which material and how much to provide to best achieve these benefits is still difficult to quantify but, as a general rule of thumb, it appears that diversity of the environment will accrue the most benefits to different aspects of the piglets' welfare. It is fast becoming recognised that an optimised gut microbiota can enhance disease resistance (Patil et al., 2020) and, as efforts are being made to reduce antibiotic use, finding alternative methods to achieve such outcomes are being sought. Increasing diet complexity, by introducing foraging opportunities to piglets pre-weaning, can increase microbial diversity and the presence of beneficial microbes, reducing the risk of post-weaning diarrhoea. Reduction of stress is also important for healthy gut development, with early-life stressful experiences demonstrated, in rodents, to lead to dysfunction of the intestinal barrier (O'Mahony et al., 2009). This is relevant for piglets, which may experience a number of acutely stressful husbandry procedures within the first few days of life.

3.2.3 Walls

The benefits of suitable walls in a farrowing environment include imparting a sense of enclosure for nest-building, providing a solid surface to support sow posture changes and thereby providing protection for piglets and also providing opportunities for both seclusion from conspecifics and interactions with them. When sows choose nest-sites they prefer a 'cul-de-sac' arrangement, providing protection on three sides, and a nest-entrance with a view, presumably for vigilance against potential threats. The degree of enclosure afforded by the nest-site will influence disturbance, and disturbance is known to adversely affect oxytocin levels (Lawrence et al., 1992). Providing sows with a supportive surface to lean against during their descent when lying reduces crushing (Baxter, 1991, Marchant et al., 2001; Damm et al., 2006) and sows prefer sloped or vertical walls to lie down against rather than farrowing rails (Damm et al., 2006). Manufacturers of outdoor farrowing huts or arks recognise that having angled walls is protective for piglets, as they prevent the sow from fully contacting the lower walls when changing posture and therefore trapping piglets. Recreating these protective elements in indoor environments requires quantification of the best dimensions/angles for the sloped walls to provide piglet escape zones, while being an attractive prospect to encourage supported lying behaviour by sows. Moustsen (2006 – cited in Pedersen et al., 2013) provided these details and alternative systems have successfully incorporated such features (Baxter et al., 2015).

While enclosure is important in the periparturient period, as lactation progresses nest-departure and re-integration of the sow and litter into the social group occur in natural conditions when the piglets are about two weeks old. Sows leave the nest to increase their foraging territory and increase feed intake during the metabolically demanding lactation phase. Piglets also benefit from this foraging activity. At the time of reintegration, prior familiarity with other group members minimises aggressive interactions that might impair fitness. Under natural or semi-natural conditions, pigs maintain established groups where aggression is regulated via an 'avoidance order', with specific behavioural patterns reducing risk of attacks by dominant individuals (Jensen, 2002). When a sow leaves the group to farrow, the longer she remains isolated, the more challenging it is for her and her litter to re-integrate. Early socialisation with other litters during lactation benefits piglets by reducing the impact of weaning (Pajor et al., 1999; Hessel et al., 2006), particularly the effects of mixing aggression, and improving piglet social skills post-weaning (Morgan et al., 2014). Alternative farrowing systems which include group housing or multisuckling allow this more natural reintegration; this was popular in the early 1980s, especially in Sweden. They are a cheaper alternative to individual housing throughout lactation and, if the benefits outlined above can be realised, they seem like a sensible choice. However, in general, group housing systems have returned poor production figures for piglet mortality (see Baxter et al., 2012 for a review). More recent research in Germany (Bohnenkamp et al., 2013), the Netherlands (van Nieuwamerongen et al., 2015) and Australia (Verdon et al., 2019; Greenwood et al., 2019) has revisited these systems, partly as a way to increase lactation length and reduce weaning stress, and thus reduce the post-weaning reliance on antibiotic interventions to maintain piglet health. However very careful management is required to ensure success of such systems (van Nieuwamerongen et al., 2014; Thomsson et al., 2016). Facilitating some form of social contact between different sows and litters in individual housing systems involves designing partitions with areas to allow fence-line contact between neighbours. Although this does not allow for full reintegration before weaning, it has been shown to benefit post-weaning outcomes such as reducing aggression upon mixing (Martin et al., 2015). Whether walls are solid or barred will also impact on dunging behaviour (Moustsen and Jensen, 2008) and the microclimate within the pen. Providing cooler (better ventilated) areas will reduce the risk of heat stress, particularly for sows during lactation (Muns et al., 2016), although thermal comfort also depends on substrate provision and floor type.

3.2.4 Flooring

Suitable flooring is one of the topics that requires significant attention in pig production systems, especially when it comes to providing for biological needs. Flooring qualities for locomotion and lying behaviour are based on many different aspects: friction, abrasiveness, hardness, surface profile and thermal properties (Lensink et al., 2013). For all livestock, floors should provide physical and thermal comfort when lying, should not lead to injury or slipping when standing and walking, but should also not be too abrasive. To promote good hygiene, they should facilitate easy cleaning, which will reduce transmission of infectious diseases (Lensink et al., 2013). There are consequently many trade-offs to consider; deciding on the optimal flooring in the farrowing and lactation environment is particularly complicated because of the diverse needs of the sows and piglets at different stages (i.e. nest-building, parturition, lactation) and because flooring interacts with many other design specifications when trying to meet those needs. For nest-building, a sow would ideally want a malleable floor to dig a hollow by rooting the ground. While this is achievable with soil under natural conditions, a proxy indoors involves provision of plentiful substrate, which means that the flooring needs to be solid to retain this substrate. This is also the case for substrate provided as enrichment material and to improve comfort by reducing the risk of development of shoulder ulcers in sows during prolonged lying bouts (Rolandsdotter et al., 2009; Rioja-Lang et al., 2018) and leg injuries in piglets during suckling (Zoric et al., 2009; Lewis et al., 2005; Mouttotou and Green, 1999). However, excreta build-up on solid floors can increase the risk of disease, and self-cleaning slatted systems tend to promote good pen hygiene (as well as reducing labour) (Rantzer and Svendsen, 2001). Good hygiene can be further promoted by clear differentiation between lying areas and areas designed for defaecation. Providing part-slatted flooring in pens helps to designate these areas, and also to provide temperature differentials as slatted floors have different thermal properties to solid floors. If given the correct cues, pigs will actively separate their dunging, resting and feeding sites (Randall et al., 1983). They will also seek corners in which to defecate (Wiegand et al., 1994), which has implications for pen design.

If plentiful substrate is not provided, soft surfaces, such as rubber matting, may allow some deformation of the surface. This reduces contact pressure and mechanical stress on the body, and can aid healing if sows have shoulder sores (Zurbrigg, 2006). Rubber mats can also reduce carpal lesions of the piglets (Courboulay et al., 2000). However, the quality and abrasiveness of the mat are important to prevent excessive wear and tear, as well as to prevent slipping. Some studies suggest rubber flooring is less slippery (Boylet et al., 2000), but others report a 'film of slurry' when mats were used in farrowing houses (Calderón Díaz et al., 2013) which will increase the slip risk for sows and impair hygiene. Flooring is a very difficult system component to perfect and, when designing a new indoor system, providing different floor types in different functional areas within the same pen is likely to be the best way to meet all biological needs.

3.2.5 Thermal comfort

The thermo-neutral zones of sows and newborn piglets are markedly different. A sow's evaporative critical temperature is the upper limit of the thermo-neutral zone and represents the temperature at which evaporative heat loss begins to increase and heat stress develops. This depends on a variety of factors associated with the sow's ability to thermoregulate and is heavily influenced by components of the housing system and ambient temperature. It has been postulated that peri-parturient sows will actively choose a nest-site based on the most thermally resistant flooring (Pedersen et al., 2006). In a study by Hunt and Petchey (1987), sows given the choice of either a concrete floor or a rubber floor with varying amounts of straw chose the mat with the greatest amount of straw. However this choice might have been dictated by comfort or substrate availability, as subsequent studies investigating nest-site choice have shown no difference between heated or room-temperature flooring with the same properties and substrate availability (Pedersen et al., 2006; Baxter et al., 2015). It is generally thought that peri-parturient sows can tolerate slightly higher temperatures than their typical thermal comfort zone during lactation of 10-20°C (Black et al., 1993) and it could be postulated that this tolerance enhances biological fitness as piglets are extremely cold sensitive. This tolerance may be limited, however, as Muns et al. (2016) found that sows in farrowing crates experiencing 25°C around parturition altered their postural behaviour relative to those kept at 20°C. They reacted to the thermal challenge with higher respiration rate, but both their rectal and udder temperatures were elevated, indicating that they were not able to compensate for the higher ambient temperature. When sows are loose-housed they have more possibility to regulate their temperature by altering their postures and choosing different areas in their environment. A study investigating three room temperatures (15°C, 20°C and 25°C) for lactating sows kept loose in pens with a partly slatted concrete floor showed that sows used the cooler slatted floor for behavioural thermoregulation by resting in this zone in between daily activity bouts (Malmkvist et al., 2012). During lactation, sows are more at risk from heat stress due to the higher metabolic activity associated with elevated feed intake and milk synthesis (Williams et al., 2013). Sows unable to thermoregulate will experience hyperthermia/heat stress, which can impact milk production and therefore piglet growth rate and survival. Thus, loose housing with different thermal zones may increase sow thermal comfort and positively affect piglet growth (e.g. Pedersen et al., 2011; Oostindjer et al., 2010).

The thermal challenge for piglets is avoiding hypothermia. At the time of birth they experience a sudden drop in ambient temperature (of approximately 15–20°C – Herpin et al., 2002). Consequently, the lower limit of the thermoneutral zone (the lower critical temperature – approximately 34°C Mount, 1968), is exceeded, resulting in chilling. It is impossible to keep the piglets in their thermoneutral zone by raising room temperature, as heat stress for the sow could be fatal. Thus, providing a suitable microclimate in a designated piglet zone is a necessity, and substrate, flooring and the degree of enclosure in the nest-site all influence the effectiveness of this. Deep bedding slows heat loss, having a thermal resistance 11 times greater than that of concrete slats and 22 times greater than solid, wet concrete flooring (Wathes and Whittemore, 2006). Mount (1967) demonstrated that piglets in contact with a concrete floor lost 40% more heat than those in contact with 2.5cm of straw. This is, in part, why straw management is such an important factor in outdoor pig production. In the absence of deep bedding, an artificially heated creep area of suitable size (Table 1) can provide an effective microclimate, provided that piglets can be attracted to use this area at an early age.

3.2.6 General system components

Although there are some advances to be made in applying nutritional strategies (see Section 3.1), it is assumed that the fundamental physiological requirements for food and water are generally well catered for. However, when sows are loose-housed, placement of feeders and drinkers can influence the spatial organisation of behaviour, especially excretory behaviour (Moustsen and Jensen, 2008; Andersen and Pedersen, 2011; Ocepek et al., 2018), and could influence farrowing location (Baxter et al., 2015). Sows will typically eat and then turn away from their feeding area to defecate, often orientating themselves so that their head faces away from their feed site (Moustsen et al., 2007; Moustsen and Jensen, 2008; Andersen and Pedersen, 2011). Incorporating barred partitions between pens in the designated dunging area, and a solid dividing wall between designated dunging and nesting sites, can improve pen hygiene (Moustsen and Jensen, 2008). However, there is large individual variation in dunging patterns in loose-housed sows (Bøe et al., 2016) and maintaining hygiene continues to be a subject needing research activity (Hansen, 2018). This is not just because good pen hygiene reduces the risk of disease transmission, but it would also encourage greater acceptance of loose-housing systems. Cleaning out farrowing pens and providing new bedding was estimated to require 33% of total daily work time for Swedish farmers on 35 commercial pig farms (Mattsson et al., 2004). The stockperson is a key contributor to animal welfare and should be considered as an important participant in system design. Stockperson intervention to maximise piglet survival has increased in the wake of superprolific breeding programs, where the number of piglets born regularly surpasses the sow's ability to rear them. Large litter size management involves handling of piglets and sows, cross-fostering, split-suckling, establishment of nurse sows, and provision of supplementary nutrition (Baxter et al., 2013, 2018). This represents a challenge in any system but could be regarded as more challenging in loose-housing conditions (Rosvold et al., 2017). Facilitating stockperson interventions while safeguarding the welfare of the sow is therefore an important element of system design. Several Danish studies have evaluated loose-housing systems and pinpoint large litter size as the main risk factor in allowing sows to be loose all the time (i.e. zero-confinement) (e.g. Hales et al., 2015). As such, there has been sustained interest in developing 'alternatives' that retain some form of restraint of the sow (i.e. temporary crating).

The increased research and development activity in alternative systems is dominated by temporary crating designs (for reviews see Baxter et al., 2018; Glencorse et al., 2019; www.freefarrowing.org). Temporary crating systems (e.g. Table 4) facilitate loose-lactation but cannot support all the biological needs of the sow, especially her functionally important nest-building behaviour (Hansen et al., 2017). These systems are therefore a compromise; they are generally the most economically attractive choice out of all the alternatives (bar outdoor production – Guy et al., 2012). They are often built on the same spatial footprint as a conventional crate system, with fully or partly slatted flooring, and maintain most of the stockperson advantages that conventional systems impart (i.e. easy and safe intervention, easy dung removal). Where pledges to phase out farrowing crates have been made (e.g. Denmark and Austria - Hansen, 2018; Heidinger et al., 2018), there has been great investment in evaluating these compromise systems, including identifying when sows should be let out of their crates post-partum. The 'critical time window' for piglet survival has been suggested as four days post-partum (Heidinger et al., 2018). It can, of course, be argued that any reduction of restraint is an improvement in animal welfare, but temporary crating systems are far from optimal for sows or piglets. Many of the benefits of allowing for more satisfaction of nest-building motivation are not realised for the sow or her piglets, and ensuring the sow is released after a short period post-partum will be based on the stockperson's discretion. There can also be spikes in piglet mortality at this time if opening protocols are sub-optimal (King et al., 2019), or if pens are operated as zero-confinement when they have been designed for temporary crating. This is repeatedly seen in studies where the same design of temporary confinement system is compared when kept in an open position all of the time versus varying lengths of confinement (e.g. Hales et al., 2014; Condous et al., 2016; Lohmeier et al., 2020). As these systems rarely have all the design features that promote good maternal behaviour (as described above), it is almost inevitable that they will fail when the sow is completely free. It is understandable that temporary crating options may be perceived by farmers as a good introduction to loose-housing, with the crate offering an 'insurance' policy to return to conventional methods if there is no uptake either by staff or the supply chain. However, even temporary crating systems represent an investment - the most basic ones (i.e. those occupying the same footprint as conventional systems) cost ~1.6 times more than farrowing crates (Guy et al., 2012) and such investment might be costly if not future-proofed for consumer demands. This was evident in the poultry industry, where the introduction of the 'enriched (or furnished) cage' to replace battery cages did not fully appreciate public perception that 'a cage is still a cage' and therefore a push for alternatives which resulted in a surge in the free range market and development of barn egg and aviary alternatives.

Thus, it is still important to determine whether optimal welfare can be attained in a farrowing and lactation system where the needs of all stakeholders are taken into account (i.e. sow, piglet, stockperson and consumer).

3.4 Case study: PigSAFE - a zero confinement farrowing and lactation system

A project commissioned by the UK government aimed to develop a suitable alternative to the farrowing crate that provides for the maximal sow and piglet welfare that can be achieved under commercial conditions. This required the re-designing of a system that could reconcile a 'triangle of needs' relating to the farmer, the sow and her litter. However, as the needs of the litter and the farmer are often closely aligned, both prioritising improved piglet survival and growth, the main conflict to be resolved lies between the sow and the farmer. This requires methods to provide the appropriate level of environmental enrichment to meet the biological 'needs' of the farrowing sow that are also consistent with good piglet survival, and other management and business constraints. A literature review was undertaken to establish/confirm the biological principles (including ethological understanding) underlying sow and piglet needs in the perinatal period. From this, a list of biological specifications (or design criteria) for zero-confinement farrowing systems was derived (as discussed in summary above and in Table 1, and in detail in Baxter et al., 2011a). A second critical review was undertaken to gather information on all known alternative farrowing and lactation systems (including outdoor systems), to identify the extent to which each system addressed biological specifications and its practical success (Baxter et al., 2012; updated in Baxter et al., 2018). From this platform, and in collaboration with representatives from industry, non-governmental organisations, biologists, economists and engineers, a prototype system was developed - PigSAFE (Piglet and Sow Alternative Farrowing Environment). This was tested on two research farms to 'model' potential implementation scenarios (i.e. conversion of existing farrowing crate accommodation (Farm A) and a new-build (Farm B)) and to rapidly de-bug obvious problems.

3.4.1 The PigSAFE design

PigSAFE is a 'designed pen'. This terminology is used for a pen that has different areas to fulfil different functions (Baxter et al., 2012, 2018) and specific pen features designed to stimulate good animal behaviour (i.e. good maternal behaviours: correct farrowing location, careful lying behaviours, appropriate interactions; good hygiene: use of provided dunging sites). The PigSAFE designs on Farm A and B are shown in Figs 1a and 2a, respectively. They have similar features, but Farm A has a narrower dunging passage and is built above a slurry pit with plastic slotted and slatted flooring. Both pens have a nest area, with solid flooring to allow provision of nesting material, and sloping walls to control the stand-to-lie posture changes of the sow and lower the risk of piglets being trapped and killed. A heated creep area has easy access from the nest for the piglets and from the passageway for stockperson inspection. A separate slatted dunging area is bounded by walls with barred panels to adjacent pens, to discourage farrowing outside the nest and allow sow-to-sow visual and oral-nasal contact. A sow feeding crate (solid-sided) is included at one side of the pen, where the sow can be locked in to allow safe inspection or treatment of the piglets, but is not wide enough to allow the sow to be locked

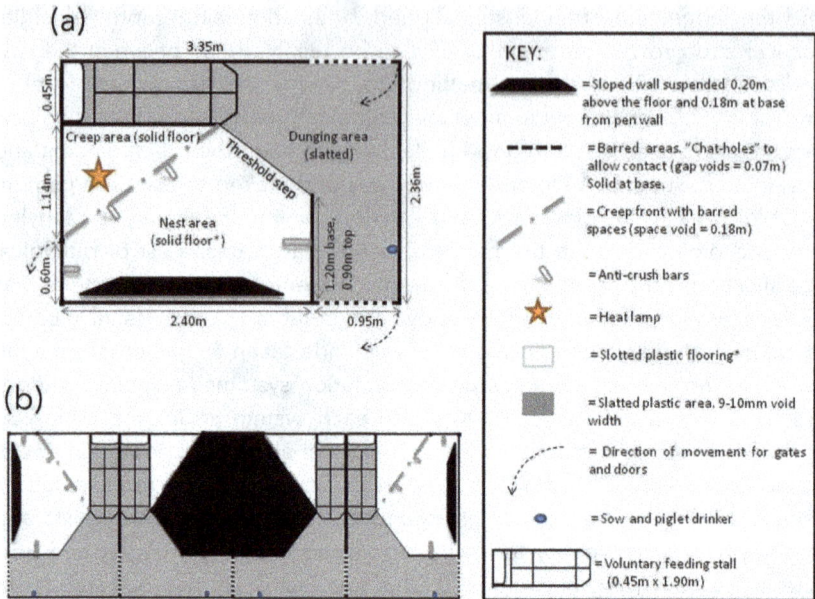

Figure 1 (a) Prototype PigSAFE pen (not for scale) and (b) Four pens were constructed in one room: in each room, two pens were provided with sound proofing material and a temporary roof fitted (height 1.2m) for the QUIET treatment.

in for farrowing. Farm B's new-build design incorporates additional dynamic features such as the ability to open up the nest area after a designated time (~7 days post-farrowing) to improve hygiene and simulate nest departure. The pen is partially slatted, with a scraper system under the metal slats in the dunging area.

The first iterations of both designs tested some gaps in the knowledge regarding the effects of a 'quiet' nest-site and quantification of substrate needed in the nest-site (Farm A – Edwards et al., 2012), as well as the effects of under-floor heating in the nest-site and quantification of nest size (Farm B – Baxter et al., 2015).

3.4.2 Initial trials

A concern when sows farrow loose is that they will not farrow in the desired location. In the PigSAFE design, that location was the nest-site where substrate was present and sloped walls provided protection and get-away sites for piglets, as well as a large heated creep. Initial trials were focussed on determining whether we could provide the correct stimuli to achieve the desired farrowing location and acceptable piglet survival (defined as comparable to farrowing crates and national herd averages).

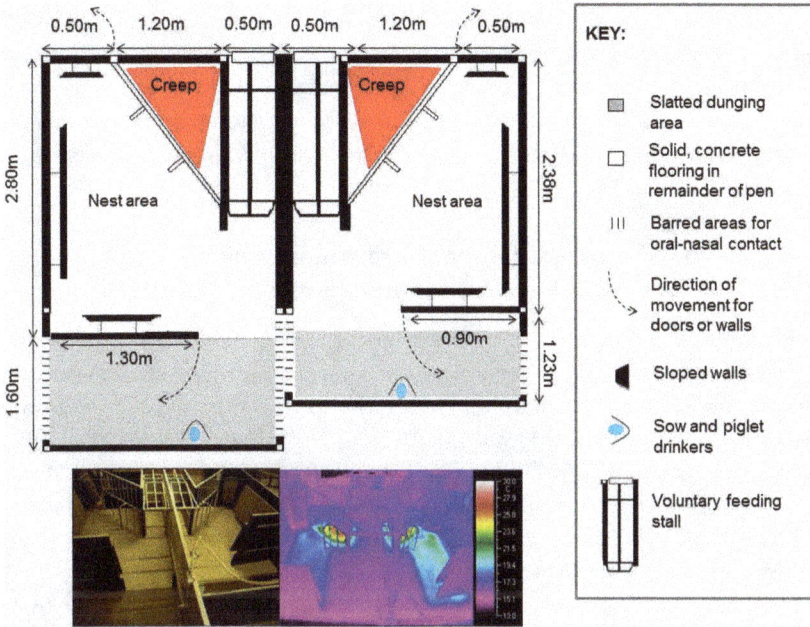

Figure 2 PigSAFE design showing LARGE (left) and SMALL (right) pen designs and thermal image showing dispersion of heat in the nest-site when under-floor heating was applied for treatments.

3.4.2.1 The influence of sound-proofing and substrate on farrowing location and performance

For the trials on Farm A, it was hypothesised that 1) 'ad libitum' substrate provided in the nest area would encourage nest-building and farrowing in the correct location, and 2) a quiet nest-site would encourage farrowing in the correct location and benefit piglet survival. The treatments were applied within the PigSAFE pen design (Fig. 1a) (overall pen size = 2.36m x 3.35m = 7.9m^2). Sound-suppression material and a roof over the nest-site were used to create a 'quiet' nest (Fig. 1b), and long-stemmed straw was provided at two levels (MAX – 4 kg and MIN – 2 kg) before farrowing until two days post-farrowing. A 2 × 2 factorial design saw 99 Large White × Landrace sows randomly assigned to quiet (i.e. quiet or control) and substrate (MAX or MIN) treatments. Linear mixed models analysed data. All the sows chose to start farrowing in the nest and 99% of all piglets were farrowed in the nest. While there were numerical trends in the expected direction, neither the degree of nest enclosure/soundproofing nor the quantity of nesting substrate supplied had a statistically significant influence on piglet survival (Live-born mortality, Quiet = 14% vs. Control = 17%, sem = 1.52, P = 0.16; Min = 17% vs. Max = 14%, sem 1.52, P = 0.11).

The overall design successfully promoted farrowing in the desired location, irrespective of substrate amount or sound-suppression. This allowed the more cost-effective and practical pen design (i.e. no covered nest and sound proofing) and substrate amount (i.e. 2 kg long-stemmed straw provided in nest-site on entry and maintained at that level until farrowing) to be taken forward for the next iterations and 'commercial' trials.

3.4.2.2 The influence of temperature and nest size on farrowing location and performance

At Farm B, two hypotheses to optimise farrowing location and improve piglet survival were tested: 1) a heated nest-site would be more attractive to the farrowing sow, and 2) greater space would improve maternal behaviour and therefore benefit piglet performance. PigSAFE (Fig. 2) was adapted: the LARGE treatment measured 9.7m^2 in total, with a nest area of 4.0m^2, and the SMALL treatment (same design but smaller) measured 7.9m^2 in total with a nest area of 3.3m^2. The nest floor was heated to either 30°C (T30) or 20°C (T20) from 48h before, until 24h after farrowing. Room temperature was kept at 18°C.

A 2 × 2 factorial design saw 88 Large White × Landrace sows randomly assigned to space and temperature treatments. Generalised linear mixed models analysed data. The overall design successfully promoted farrowing in the desired location, irrespective of nest size and floor temperature (97% of all piglets born were born in the nest). There was no significant effect of floor temperature on performance. However, space influenced mortality, with significantly greater live-born mortality when sows were afforded a larger farrowing space (LARGE=18.1% vs. SMALL=10.9% P=0.028). Behavioural analysis suggested that the greater mortality was a result of a larger nest area in which the sow could lie down unsupported (full results of this study can be found in Baxter et al., 2015). These results suggest the larger nest size was less protective for the piglets and thus a smaller nest would be recommended. The pen features with nesting substrate provided enough stimuli, regardless of floor temperature, to attract sows into the nest.

Conclusion from the initial trials: the results indicated that the PigSAFE pen provided the correct stimuli (enclosed nest-site, substrate) to attract the sow to farrow in the desired location and that piglet survival levels were acceptable at both sites within the constraints of the experiment. The final prototypes would adopt the best designs at both sites (which were also the most cost-effective) to run 'commercially' using farm rather than research staff.

3.4.3 PigSAFE 'commercial' trials and performance

For this phase, farm staff at each site were responsible for operating both PigSAFE and farrowing crates in parallel, given minimal experimental constraints.

The optimised designs were used at each site and only basic instructions were given, including the best practice for substrate provision based on initial trials. Data presented are on pig performance.

3.4.3.1 Piglet survival

Combined data from the two sites (304 sows, n=164 in Crates n=140 in PigSAFE, approximately 10% gilts) are shown in Table 2 against the national average and top third data at the time of the trial.

When the data at each farm were looked at separately, there were interesting trends suggesting farm differences. At Farm A, the PigSAFE system performed very well, with no significant difference compared to crates (Fig. 3) and both systems performing better than the industry top third. At Farm B, although there were no significant differences in pig performance, mortality in the PigSAFE pen was closer to the industry indoor average.

At Farm B, pig performance in PigSAFE pens showed an improvement with every batch farrowed (Fig. 4), highlighting that stock-person training is an important aspect when adopting a new system. In contrast with Farm A, where the manager had previous experience with loose-farrowing outdoor sows, the manager at Farm B had previously worked with sows in crates for over 25 years and thus was unfamiliar with loose farrowing and lactating sows. The learning curve is indicative of potential scenarios when implementing a new system on a commercial farm where staff have previously had little or no experience of animals farrowing and lactating while loose.

Table 2 Performance of PigSAFE pens and farrowing crates for the period up to August 2011, with national average and top third farrowing crate data for comparison (AHDB Pork, 2010[1])

	Total born (BA+BD)	Total weaned	%Total mortality	%Live-born mortality	%Stillborn mortality*
PigSAFE	13.0	10.7[a]	13.3[b]	9.6[c]	4.1[b]
Farrowing crates	12.8	10.8[a]	12.0[b]	9.0[c]	3.4[b]
National indoor average	12.3	10.1	17.6	12.3	6.1
National indoor top third	13.4	11.2	16.5	10.3	5.6

[a]Adjusted for net fostering; [b]This figure includes both stillborn piglets and those dying before weaning and is adjusted for litter size; [c]Adjusted for litter size post fostering. BA = Born alive, BD = Born dead/ Stillborn. *Stillborn% = percentage of total litter size that were born dead.
[1] https://porktools.ahdb.org.uk/prices-stats/costings-herd-performance/indoor-breeding-herd/.

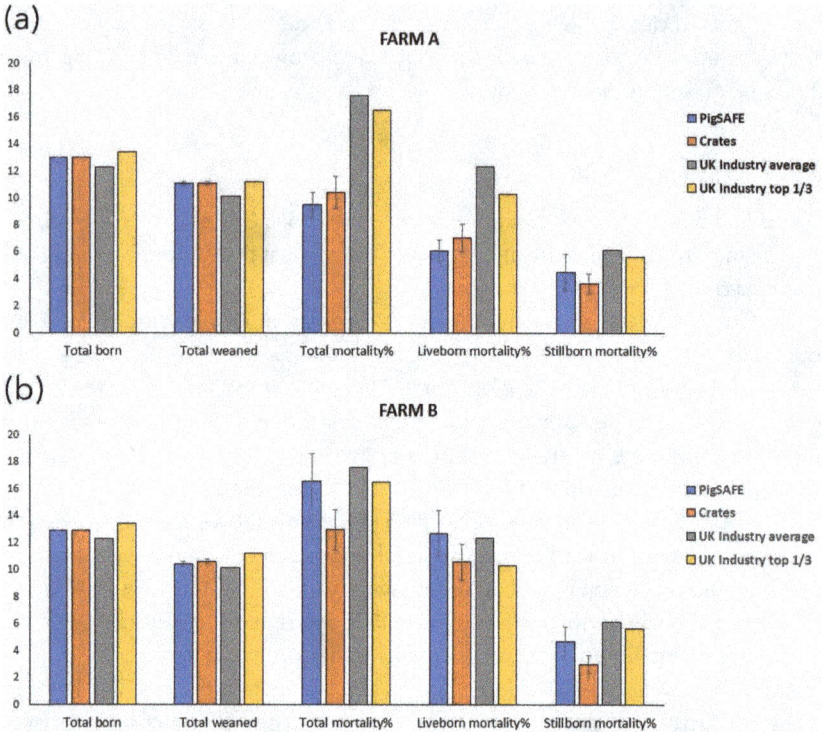

Figure 3 Performance at Farm A (a) and Farm B (b). Descriptive comparison against industry average and top third at the time (AHDB Pork, 2010, https://porktools.ahdb.org.uk/prices-stats/costings-herd-performance/indoor-breeding-herd/). Mortality data are adjusted for net fostering (similar in both systems), parity and litter size.

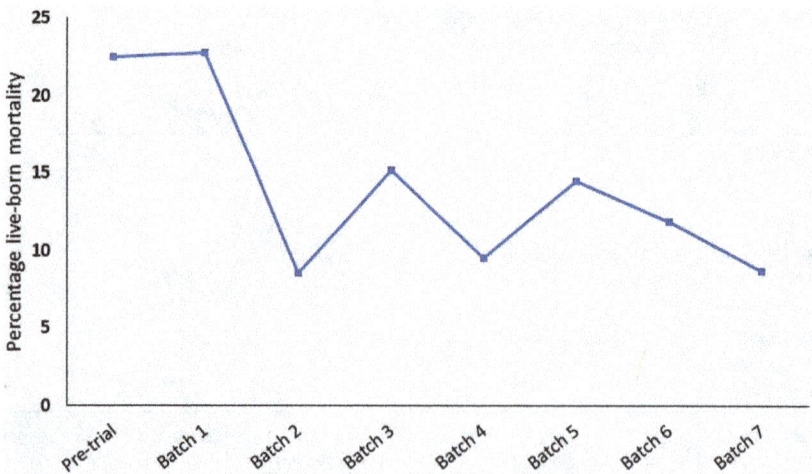

Figure 4 Effect of staff experience on live-born mortality in the PigSAFE system at Farm B.

3.4.3.2 Sow and piglet weight and condition

Body weight (kg) and back-fat depth were measured on sows pre-farrowing and post-weaning, and piglets were weighed at weaning. Table 3 shows that although there were no significant differences between housing systems at Farm A, at Farm B piglets weaned from PigSAFE pens were significantly heavier than those from crates (average individual weight 8.8 kg vs. 8.5 kg).

At Farm B there was a tendency for sows housed in PigSAFE pens to eat more than those housed in crates (7.27 kg vs. 6.44 kg per sow per day $F_{1,6}$=3.45 P=0.088). At Farm A there were no significant differences in feed intake between farrowing systems (P=0.621). It is possible that the difference in feed intake is only seen at Farm B as a result of potentially beneficial effects of the building in which the PigSAFE pens were built. The 'new build' scenario implemented at Farm B involved a large building shell with high roof space, and thus a more airy environment for the sows. It is possible that the higher piglet weaning weights seen in PigSAFE at Farm B relate to this sow feed intake, but it could also be that the more open pens (at 7 days post-farrowing the dividing wall opened up) improved udder access and milk let-down. This has been seen in other work demonstrating the benefits of space on suckling success as a result of easier udder access (Pedersen et al., 2011). These performance advantages are important, as the capital investment required to install a PigSAFE pen has been shown to be more than a conventional crate (Guy et al., 2012).

Conclusions from commercial-type trials: Results from the commercial scale-up phase showed no significant differences in performance between PigSAFE pens and conventional crates within farms. The most promising results from Farm A showed that the PigSAFE system can perform better than the industry top third producers. A key aspect of any new system is the ability and willingness of stockpersons to adapt to a new system and this effect was clearly

Table 3 Influence of farrowing accommodation on sow body condition and piglet weaning weight (adjusted for litter size, weaned litter size and weaning age). Sow weight loss includes the birth of the litter. For statistical purposes, systems within sites were compared.

| | Sow condition | | Litter condition |
	Weight loss (kg)	Back-fat loss (mm)	Average litter weaned weight (kg)
Farm A PigSAFE	38.81	4.18	78.05
Farm A Farrowing crates	38.54	4.28	76.14
Farm B PigSAFE	27.68	4.45	91.33[a]
Farm B Farrowing crates	29.71	4.18	87.77[b]

Subscripts with different letters indicate figures are significantly different at $P < 0.05$.

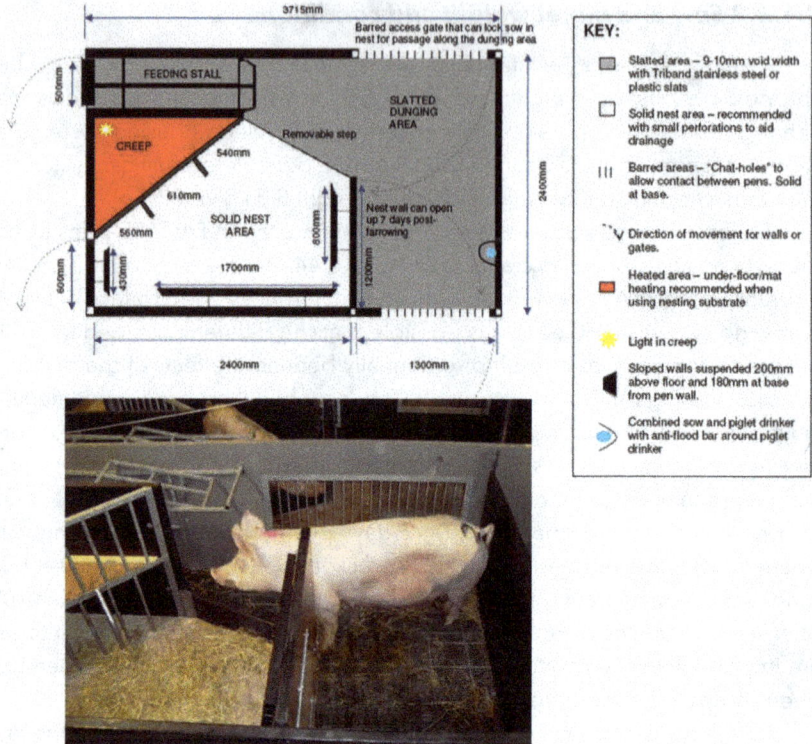

Figure 5 Recommended design and dimensions for building a PigSAFE pen.

seen at Farm B. There is also some indication of benefits in weaning weights and sow feed intake when using PigSAFE pens compared with conventional crates. The recommended best design as a result of the research is shown in Fig. 5, and full details of all components can be found at 'www.freefarrowing .org'.

Throughout the project it was emphasised that the detail of the design is very important in its performance and therefore any deviation from the recommended details is done at the farmer's own risk.

3.5 Commercial developments in alternative farrowing and lactation systems

Since the publication of the initial review manuscripts (Baxter et al., 2011a, 2012), there has been a great deal of research and industry activity in the development of alternative systems (Table 4). These have included large national studies in Austria (Pro-SAU) and Denmark (SEGES showroom) and plans for reform to phase out farrowing crates in these countries. Thus Baxter

Table 4 Examples of different farrowing and lactation systems (Specific information on alternative systems can be found at www.freefarrowing.org)

System		Descriptions
Crates	Conventional	Crate	Tubular metal bars run horizontally along the length of the crate, with additional bars positioned above the sow to prevent escape via climbing or jumping. Crates are fixed width or can be adjusted at the back for larger sows. Space available to sow: 0.90–1.23 m². Total space available to piglets: ~3.5–4.0 m². A water and food trough is situated at the front with piglet water drinker within pen. Flooring: fully or part-slatted with pens usually having a solid floor or mat for the piglet resting area, which includes a heat source.	Fixed-width farrowing crate with front covered and heated creep, solid floors with cast-iron slatted area at rear for waste removal. Photo courtesy of E. M. Baxter.

Adjustable farrowing crate with capability to widen but prevent turning. Corner, heated creep, partially slatted floors (concrete/cast-iron). Photo courtesy of V. A. Moustsen. |

(Continued)

Table 4 (*Continued*)

System		Descriptions	
Temporary crating	Hinged/ swing-side for example, 360° MPP, Combi-flex, Combi-flex hexel, Winged	Farrowing crate structure within pen as in conventional systems. Spatial footprint and flooring same as conventional system. Sow is crated on entry to farrowing house until approximately 5–7 days post-farrowing. Crate can be opened up or swung-open for lactation allowing the sow to turn around. An additional heated and solid area (typically a heat-mat) inaccessible to the sow is provided for the piglets.	360° MPP temporary crate in 'open' (left) and 'closed' (right) position. Crate typically opened after 4–7 days. Heat mat for piglets on fully slatted plastic floor. Photo courtesy of E. M. Baxter. 'Winged' temporary crating option. Crate opens up at the back to allow the sow to reverse out and turn in the dunging area. Photo courtesy of V. A. Moustsen.

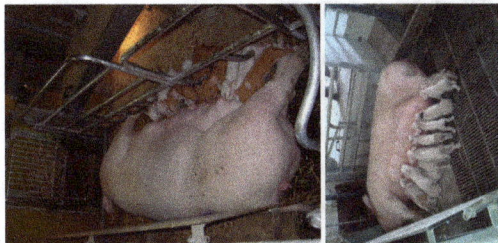

SWAP in closed and open positions. Photo courtesy of V. A. Moustsen.

Simple pen on the same spatial footprint as a conventional farrowing crate. Fully slatted flooring. Photo courtesy of V. A. Moustsen.

Pens		
	SWAP	The SWAP system (Sow Welfare And Piglet Protection) is designed to allow nest-building and then confinement post-farrowing until approximately four days post-farrowing (Moustsen et al., 2013). Sow should be able to turn around for nesting and lactation but crated for farrowing and immediately post-partum. Larger footprint than the crate – 6 m² built on part-slatted flooring. Separate large corner heated and covered area for piglets.
	Simple	For example, simple pen, mushroom pen, sloped floor	Same spatial footprint as conventional crate. No crate. Sow able to turn around at all times. Fully or part-slatted flooring. Possibility of modified floor design. Separate heated area for piglets.

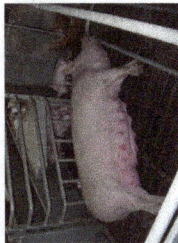

(Continued)

Table 4 Examples of different farrowing and lactation systems (Specific information on alternative systems can be found at www.freefarrowing.org) (Continued)

System		Descriptions	
Designed pen	For example, Comfort Sow, Danish FF, FATs, PigSAFE, WelCon, Werribee	Zero-confinement. Sow can turn around. Defined areas provided in the pen for feeding, dunging and lying/nesting. Size varies (5–8.5 m²). Additional pen 'furniture' such as rails or sloped walls to assist sow posture changes and protect piglets. Solid or part-slatted floor with generous lying area to provide substrate for nest-building. A separate, heated creep area for the piglets, inaccessible to the sow is provided. Some systems have the potential for separating the sow from piglets to allow farmers to perform husbandry procedures safely.	Danish Free Farrower Zero confinement pen with part-slatted flooring and a corner creep. Photo courtesy of V. A. Moustsen. PigSAFE zero confinement pen with part slatted flooring, corner creep and separate feeding, nesting and dunging areas. Photo courtesy of E. M. Baxter.

| Group | Zero-confinement group | Separate or free-access nests then group | Systems allow sows and litters to mix before weaning, typically 10–21 days into lactation. Based on multi-suckling accommodation (e.g. Swedish systems – Ljungström, Thorstensson). Variable but large amounts of space for both sows and piglets.

For farrowing, sows are initially individually housed in single pens, but are integrated with their litter into groups in larger multi-suckling pens between 10–21 days post-farrowing Alternatively, sows are already grouped prior to farrowing and have free access to individual nest boxes for farrowing. | 
Typical Swedish multi-suckling system with deep-straw bedding. Photo courtesy of J. H. Pedersen.

Group housing system for lactating sows at Swine Innovation Center Sterksel in the Netherlands.
Photo courtesy of C. van der Peet-Schwering, Wageningen University and Research Centre.

(Continued) |

Table 4 Examples of different farrowing and lactation systems (Specific information on alternative systems can be found at www.freefarrowing.org) (Continued)

System		Descriptions	
Kennels	Temporary crating then group	Sows are conventionally crated and then grouped 10-14 days post-partum to allow litter integration.	
	Crated then grouped Kennel with run, for example, 'Solari'	An outside space is intended for dunging and feeding by the individually housed sow, with an indoor kennel for farrowing. Floors are solid to facilitate provision of deep substrate. A heated, creep space may be provided within the kennel for the piglets.	Outdoor kennel and run. Photo courtesy of E. M. Baxter.
Outdoor	Outdoor	These are systems with low capital investment and running costs, where sows and their piglets are housed individually, outdoors in farrowing arks or huts, with access to individual or group paddocks. There are different ark and hut designs available and described in detail elsewhere (e.g. Honeyman et al., 1998; Baxter et al., 2009).	Outdoor farrowing huts and paddocks. Photos courtesy of S.-L. A. Schild (top) and E. M. Baxter (bottom).

et al. (2018) updated their earlier scientific review to include new information in the state-of-the-art. In addition, Baxter and Edwards developed a website to provide a more farmer-friendly interface for those interested in building alternative systems (www.freefarrowing.org).

Many of the alternative systems available commercially are temporary crating systems, as farmers seek to retain greater control over the sow's movements, particularly around the first few days post-partum when interventions to promote piglet survival and piglet husbandry procedures are performed. Farmers have also said that they want an 'insurance policy' of being able to resort to the conventional crated system if they do not succeed in adopting free farrowing or loose lactation, either because of stockperson reluctance, performance problems or a failure in the market to recognise the additional financial requirements of higher welfare systems. Although there has been a lot of work on pen design, less attention has been given to optimising management and genetic selection of the best animals for loose indoor systems. It is likely that all three 'Ps' (Pens-People-Pigs) will need to be optimised to achieve consistent performance and reduce the barriers to uptake.

4 Conclusions

Determining the basic needs of all actors (animals, farmers) that are involved in a system is a key starting point for designing higher welfare, sustainable alternatives to conventional farrowing crate systems that are known to impose welfare challenges for both sows and piglets.

Allowing the animals to be more active in the control of their environment is a positive aspect of animal welfare that appears to be an important part of their behavioural needs. When given the chance to be an active participant in her own environment, the sow will fulfil her own needs but also contribute to the goals of others. Maximising piglet survival increases her biological fitness and also satisfies the main needs of the piglets and farmer. The current challenge is what happens when the biological limits of the sow have been exceeded and she can no longer fully control the outcomes of all her offspring without the contribution of another actor. Designing systems that can allow stockperson interventions without imposing any further constraints on sow welfare is an important goal for research and development.

5 Future trends in research

This chapter has concentrated on the managerial and environmental interventions required to optimise animal welfare around farrowing. Research on the development of alternative farrowing and lactation systems is reasonably mature; however, greater innovation is required to optimise system components. For example, designing suitable flooring and waste management systems that allow the provision of environmental enrichment without compromising hygiene is still a challenge.

Optimising the biological components of farrowing and lactation systems will involve breeding for improved maternal behaviour (Grandison, 2005; Gäde et al., 2008; Baxter et al., 2011b) and investigating strategies for breeding a more robust piglet. Both have potential to improve piglet survival and therefore the potential to promote greater confidence with loose housing. However, breeding strategies seeking short-term gains in production, such as selecting for increased numbers of piglets born beyond the biological limits of the sow to rear them, threaten welfare. They also challenge system design to incorporate features that help cater for the increasingly specialised needs of piglets born into large litters. Finally, a social science element meriting further research is understanding the barriers to uptake of alternative systems, including stockperson behaviour and willingness to change. Greater attention to all these aspects should improve sow and piglet welfare while addressing farmer concerns.

6 Where to look for further information

In addition to the peer-reviewed literature provided in the reference section, web-based information can be found detailing practical ways to optimise the farrowing and lactation environment.

- www.freefarrowing.org.
- http://practicalpig.ahdb.org.uk/.

7 References

Ahlström, S., Jarvis, S. and Lawrence, A. B. 2002. Savaging gilts are more restless and more responsive to piglets during the expulsive phase of parturition. *Applied Animal Behaviour Science* 76(1), 83-91.

Alexopoulos, J. G., Lines, D. S., Hallett, S. and Plush, K. J. 2018. A review of success factors for piglet fostering in lactation. *Animals* 8(3), 38.

Algers, B. and Uvnäs-Moberg, K. 2007. Maternal behavior in pigs. *Hormones and Behavior* 52(1), 78-85.

Amdi, C., Hansen, C. F., Krogh, U., Oksbjerg, N. and Theil, P. K. 2013a. Less brain sparing occurs in severe intrauterine growth-restricted piglets born to sows fed palm fatty acid distillate. In: *Manipulating Pig Production XIV*. APSA Biennial Conference, Melbourne, Victoria, Australia, 24th to 27th November, p. 125.

Amdi, C., Krogh, U., Flummer, C., Oksbjerg, N., Hansen, C. F. and Theil, P. K. 2013b. Intrauterine growth restricted piglets defined by their head shape ingest insufficient amounts of colostrum. *Journal of Animal Science* 91(12), 5605-5613.

Andersen, H. M. L. and Pedersen, L. J. 2011. The effect of feed trough position on choice of defecation area in farrowing pens by loose sows. *Applied Animal Behaviour Science* 131(1-2), 48-52.

Andersen, I. L., Vasdal, G. and Pedersen, L. J. 2014. Nest building and posture changes and activity budget of gilts housed in pens and crates. *Applied Animal Behaviour Science* 159, 29-33.

Arey, D. S. 1992. Straw and food as reinforcers for prepartal sows. *Applied Animal Behaviour Science* 33(2-3), 217-226.

Balzani, A., Cordell, H. J., Sutcliffe, E. and Edwards, S. A. 2016. Heritability of udder morphology and colostrum quality traits in swine. *Journal of Animal Science* 94(9), 3636-3644.

Bandrick, M., Pieters, M., Pijoan, C., Baidoo, S. K. and Molitor, T. W. 2011. Effect of cross-fostering on transfer of maternal immunity to Mycoplasma hyopneumoniae to piglets. *Veterinary Record* 168(4), 100.

Barnett, J. L., Hemsworth, P. H., Cronin, G. M., Jongman, E. C. and Hutson, G. D. 2001. A review of the welfare issues for sows and piglets in relation to housing. *Australian Journal of Agricultural Research* 52(1), 1-28.

Baxter, E. M. and Edwards, S. A. 2018. Piglet mortality and morbidity: inevitable or unacceptable? In: Špinka, M. (Ed.), *Advances in Pig Welfare*, pp. 73-100. Woodhead Publishing, Cambridge, UK.

Baxter, E. M., Jarvis, S., D'eath, R. B., Ross, D. W., Robson, S. K., Farish, M., Nevison, I. M., Lawrence, A. B. and Edwards, S. A. 2008. Investigating the behavioural and physiological indicators of neonatal survival in pigs. *Theriogenology* 69(6), 773-783.

Baxter, E. M., Jarvis, S., Sherwood, L., Robson, S. K., Ormandy, E., Farish, M., Smurthwaite, K. M., Roehe, R., Lawrence, A. B. and Edwards, S. A. 2009. Indicators of piglet survival in an outdoor farrowing system. *Livestock Science* 124(1-3), 266-276.

Baxter, E. M., Lawrence, A. B. and Edwards, S. A. 2011a. Alternative farrowing systems: design criteria for farrowing systems based on the biological needs of sows and piglets. *Animal: An International Journal of Animal Bioscience* 5(4), 580-600.

Baxter, E. M., Jarvis, S., Sherwood, L., Farish, M., Roehe, R., Lawrence, A. B. and Edwards, S. A. 2011b. Genetic and environmental effects on piglet survival and maternal behaviour of the farrowing sow. *Applied Animal Behaviour Science* 130(1-2), 28-41.

Baxter, E. M., Lawrence, A. B. and Edwards, S. A. 2012. Alternative farrowing accommodation: welfare and economic aspects of existing farrowing and lactation systems for pigs. *Animal* 6(1), 96-117.

Baxter, E. M., Rutherford, K. M. D., d'Eath, R. B., Arnott, G., Turner, S. P., Sandøe, P., Moustsen, V. A., Thorup, F., Edwards, S. A. and Lawrence, A. B. 2013. The welfare implications of large litter size in the domestic pig II: management factors. *Animal Welfare* 22(2), 219-238.

Baxter, E. M., Adeleye, O. O., Jack, M. C., Farish, M., Ison, S. H. and Edwards, S. A. 2015. Achieving optimum performance in a loose-housed farrowing system for sows: the effects of space and temperature. *Applied Animal Behaviour Science* 169, 9-16.

Baxter, E. M., Andersen, I. L. and Edwards, S. A. 2018. Sow welfare in the farrowing crate and alternatives. In: Špinka, Marek (Ed.), *Advances in Pig Welfare*, pp. 27-72. Woodhead Publishing, Cambridge, UK.

Baxter, M. R. and Schwaller, C. E. 1983. Space requirements for sows in confinement. In: Baxter, S. H., Baxter, M. R. and MacCormack, J. (Eds), *Farm Animal Housing and Welfare*, pp. 181-195. Martinus Nijhoff Publishers, The Hague, The Netherlands.

Baxter, M. R. 1991. The Freedom farrowing system. *Farm Building Progress* 104, 9-15.

Björkman, S., Oliviero, C., Rajala-Schultz, P. J., Soede, N. M. and Peltoniemi, O. A. T. 2017. The effect of litter size, parity and farrowing duration on placenta expulsion and retention in sows. *Theriogenology* 92, 36–44.

Black, J. L., Mullan, B. P., Lorschy, M. L. and Giles, L. R. 1993. Lactation in the sow during heat stress. *Livestock Production Science* 35(1–2), 153–170.

Bøe, K. E., Kvaal, I., Hall, E. J. S. and Cronin, G. M. 2016. Individual differences in dunging patterns in loose-housed lactating sows. *Acta Agriculturae Scandinavica, Section A– Animal Science* 66(4), 221–230.

Bohnenkamp, A. L., Traulsen, I., Meyer, C., Müller, K. and Krieter, J. 2013. Comparison of growth performance and agonistic interaction in weaned piglets of different weight classes from farrowing systems with group or single housing. *Animal: An International Journal of Animal Bioscience* 7(2), 309–315.

Bolhuis, J. E., Raats-van den Boogaard, A. M. E., Hoofs, A. I. J. and Soede, N. M. 2018. Effects of loose housing and the provision of alternative nesting material on peri-partum sow behaviour and piglet survival. *Applied Animal Behaviour Science* 202, 28–33.

Bonde, M. 2008. Prevalence of decubital shoulder lesions in Danish sow herds. Internal Report 12, University of Aarhus, Faculty of Agricultural Sciences, p. 8.

Bontempo, V. and Jiang, X. R. 2015. Feeding various fat levels in sows: effects on immune status and performance of sows and piglets. In: Farmer, C. (Ed.), *The Gestating and Lactating Sow*, pp. 357–375. Wageningen Academic Publishers, Netherlands.

Boogaard, B. K., Boekhorst, L. J. S., Oosting, S. J. and Sørensen, J. T. 2011. Socio-cultural sustainability of pig production: citizen perceptions in the Netherlands and Denmark. *Livestock Science* 140(1–3), 189–200.

Bos, B. and Koerkamp, P. G. 2009. Synthesising needs in system innovation through structured dresign: a methodical outline of the role of needs in reflexive interactive design (RIO). Sustainable agriculture and food chains in peri-urban areas. In: Poppe, K. J., Termeer, K. and Slingerland, M. (Eds), *Transitions Towards Sustainable Agriculture and Food Chains in Peri-Urban Areas*, Chapter 12p. 219–237. Wageningen Academic Publishers.

Boyle, L. A., Leonard, F. C., Lynch, P. B. and Brophy, P. 2002. Effect of gestation housing on behaviour and skin lesions of sows in farrowing crates. *Applied Animal Behaviour Science* 76(2), 119–134.

Boyle, L. A., Regan, D., Leonard, F. C., Lynch, P. B. and Brophy, P. 2000. The effect of mats on the welfare of sows and piglets in the farrowing house. *Animal Welfare* 9(1), 39–48.

Brown, S. M., Klaffenböck, M., Nevison, I. M. and Lawrence, A. B. 2015. Evidence for litter differences in play behaviour in pre-weaned pigs. *Applied Animal Behaviour Science* 172, 17–25.

Buller, H., Blokhuis, H., Jensen, P. and Keeling, L. 2018. Towards farm animal welfare and sustainability. *Animals* 8(6), 81.

Cabrera, R. A., Boyd, R. D., Jungst, S. B., Wilson, E. R., Johnston, M. E., Vignes, J. L. and Odle, J. 2010. Impact of lactation length and piglet weaning weight on long-term growth and viability of progeny. *Journal of Animal Science* 88(7), 2265–2276.

Calderón Díaz, J. A., Fahey, A. G., KilBride, A. L., Green, L. E. and Boyle, L. A. 2013. Longitudinal study of the effect of rubber slat mats on locomotory ability, body, limb and claw lesions, and dirtiness of group housed sows. *Journal of Animal Science* 91(8), 3940–3954.

Calderón Díaz, J. A., Diana, A., Boyle, L. A., Leonard, F. C., McElroy, M., McGettrick, S., Moriarty, J. and García Manzanilla, E. 2017. Delaying pigs from the normal production flow is associated with health problems and poorer performance. *Porcine Health Management* 3(1), 13.

Chen, J. and Lobo, A. 2012. Organic food products in China: determinants of consumers' purchase intentions. *The International Review of Retail, Distribution and Consumer Research* 22(3), 293-314.

Condous, P. C., Plush, K. J., Tilbrook, A. J. and Van Wettere, W. H. E. J. 2016. Reducing sow confinement during farrowing and in early lactation increases piglet mortality. *Journal of Animal Science* 94(7), 3022-3029.

Council Directive 2008/120/EC of 18 December 2008 laying down minimum standards for the protection of pigs.

Courboulay, V., le Roux, A., Collin, F., Dutertre, C. and Rousseau, P. 2000. Incidence du type de sol en maternite sue le confort de la truie et des porcelets. *Journees Recherche Porcine en France* 32, 115-122.

Cronin, G. M., Barnett, J. L., Hodge, F. M., Smith, J. A. and McCallum, T. H. 1991. The welfare of pigs in two farrowing/lactation environments: cortisol responses of sows. *Applied Animal Behaviour Science* 32(2-3), 117-127.

Cronin, G. M., Smith, J. A., Hodge, F. M. and Hemsworth, P. H. 1994. The behaviour of primiparous sows around farrowing in response to restraint and straw bedding. *Applied Animal Behaviour Science* 39(3-4), 269-280.

Damm, B. I., Vestergaard, K. S., Schroder-Petersen, D. L. and Ladewig, J. 2000. The effect of branches on prepartum nest building in gilts with access to straw. *Applied Animal Behaviour Science* 69(2), 113-124.

Damm, B. I., Moustsen, V., Jørgensen, E., Pedersen, L. J., Heiskanen, T. and Forkman, B. 2006. Sow preferences for walls to lean against when lying down. *Applied Animal Behaviour Science* 99(1-2), 53-63.

D'Eath, R. B., Tolkamp, B. J., Kyriazakis, I. and Lawrence, A. B. 2009. 'Freedom from hunger' and preventing obesity: the animal welfare implications of reducing food quantity or quality. *Animal Behaviour* 77(2), 275-288.

de Barcellos, M. D., Grunert, K. G., Zhou, Y., Verbeke, W., Perez-Cueto, F. J. A. and Krystallis, A. 2013. Consumer attitudes to different pig production systems: a study from mainland China. *Agriculture and Human Values* 30(3), 443-455.

Decaluwé, R., Maes, D., Wuyts, B., Cools, A., Piepers, S. and Janssens, G. P. J. 2014. Piglets' Colostrum intake associates with daily weight gain and survival until weaning. *Livestock Science* 162, 185-192.

Declerck, I., Dewulf, J., Piepers, S., Decaluwé, R. and Maes, D. 2015. Sow and litter factors influencing colostrum yield and nutritional composition. *Journal of Animal Science* 93(3), 1309-1317.

De Jonge, F. H., Bokkers, E. A. M., Schouten, W. G. P. and Helmond, F. A. 1996. Rearing piglets in a poor environment: developmental aspects of social stress in pigs. *Physiology and Behavior* 60(2), 389-396.

Devillers, N., Le Dividich, J. and Prunier, A. 2011. Influence of colostrum intake on piglet survival and immunity. *Animal: An International Journal of Animal Bioscience* 5(10), 1605-1612.

De Vos, M., Huygelen, V., Willemen, S., Fransen, E., Casteleyn, C., Van Cruchten, S., Michiels, J. and Van Ginneken, C. 2014. Artificial rearing of piglets: effects on small intestinal morphology and digestion capacity. *Livestock Science* 159, 165-173.

Donovan, T. S. and Dritz, S. S. 2000. Effect of split nursing on variation in pig growth from birth to weaning. *Journal of the American Veterinary Medical Association* 217(1), 79-81.

Douglas, S. L., Edwards, S. A., Sutcliffe, E., Knap, P. W. and Kyriazakis, I. 2013. Identification of risk factors associated with poor lifetime growth performance in pigs. *Journal of Animal Science* 91(9), 4123-4132.

Douglas, S. L., Edwards, S. A. and Kyriazakis, I. 2014. Management strategies to improve the performance of low birth weight pigs to weaning and their long term consequences. *Journal of Animal Science* 92(5), 2280-2288.

Drake, A., Fraser, D. and Weary, D. M. 2008. Parent-offspring resource allocation in domestic pigs. *Behavioral Ecology and Sociobiology* 62(3), 309-319.

D'Silva, J. and Turner, J. (Eds) 2012. *Animals, Ethics and Trade: The Challenge of Animal Sentience*. Earthscan, London, UK.

Duncan, I. J. 2005. Science-based assessment of animal welfare: farm animals. *Revue Scientifique et Technique* 24(2), 483-492.

Edwards, S. A. 2002. Perinatal mortality in the pig: environmental or physiological solutions? *Livestock Production Science* 78(1), 3-12.

Edwards, S. A., Brett, M., Seddon, Y. M., Ross, D. and Baxter, E. M. 2012. Evaluation of nest design and nesting substrate options for the PigSAFE free farrowing pen. *Proceedings of the British Society of Animal Science*, 7.

Edwards, S. A., Matheson, S. M. and Baxter, E. M. 2019a. Genetic influences on intra-uterine growth retardation of piglet and management interventions for low birth weight piglets. In: González, E. R. (Ed.), *Nutrition of Hyperprolific Sows*, pp. 141-166. Novus International, Inc., Spain.

Edwards, L. E., Plush, K. J., Ralph, C. R., Morrison, R. S., Acharya, R. Y. and Doyle, R. E. 2019b. Enrichment with Lucerne hay improves sow maternal behaviour and improves piglet survival. *Animals* 9(8), 558.

Engelsmann, M. N., Hansen, C. F., Nielsen, M. N., Kristensen, A. R. and Amdi, C. 2019. Glucose injections at birth, warmth and placing at a nurse sow improve the growth of IUGR piglets. *Animals* 9(8), 519.

Eurobarometer 2016. *Special Eurobarometer* 442./Wave EB84.4 – TNS Opinion & Social. Attitudes of Europeans towards Animal Welfare, Available at: http://ec.europa.eu/COMMFrontOffice/PublicOpinion/index.cfm/ResultDoc/download/Document.

Feldpausch, J. A., Jourquin, J., Bergstrom, J. R., Bargen, J. L., Bokenkroger, C. D., Davis, D. L., Gonzalez, J. M., Nelssen, J. L., Puls, C. L., Trout, W. E. and Ritter, M. J. 2019. Birth weight threshold for identifying piglets at risk for preweaning mortality. *Translational Animal Science* 3(2), 633-640.

Feyera, T., Pedersen, T. F., Krogh, U., Foldager, L. and Theil, P. K. 2018. Impact of sow energy status during farrowing on farrowing kinetics, frequency of stillborn piglets, and farrowing assistance. *Journal of Animal Science* 96(6), 2320-2331.

Fox, C., Merali, Z. and Harrison, C. 2006. Therapeutic and protective effect of environmental enrichment against psychogenic and neurogenic stress. *Behavioural Brain Research* 175(1), 1-8.

Gäde, S., Bennewitz, J., Kirchner, K., Looft, H., Knap, P. W., Thaller, G. and Kalm, E. 2008. Genetic parameters for maternal behaviour traits in sows. *Livestock Science* 114(1), 31-41.

Gaskin, H. R. and Kelly, K. W. 1995. Immunology and neonatal Mortality. In: Varley, M. A. (Ed.), *The Neonatal Pig: Development and Survival*, pp. 39-56. CAB International, UK.

Glencorse, D., Plush, K., Hazel, S., D'Souza, D. and Hebart, M. 2019. Impact of non-confinement accommodation on farrowing performance: A systematic review and meta-analysis of farrowing crates Versus pens. *Animals* 9(11), 957.

Gonyou, H. W., Brumm, M. C., Bush, E., Deen, J., Edwards, S. A., Fangman, T., McGlone, J. J., Meunier-Salaun, M., Morrison, R. B., Spoolder, H., Sundberg, P. L. and Johnson, A. K. 2006. Application of broken-line analysis to assess floor space requirements of nursery and grower-finisher pigs expressed on an allometric basis. *Journal of Animal Science* 84(1), 229-235.

Grandinson, K. 2005. Genetic background of maternal behaviour and its relation to offspring survival. *Livestock Production Science* 93(1), 43-50.

Greenwood, E. C., van Dissel, J., Rayner, J., Hughes, P. E. and van Wettere, W. H. E. J. 2019. Mixing sows into alternative lactation housing affects sow aggression at mixing, future reproduction and piglet injury, with marked differences between multisuckle and sow separation systems. *Animals* 9(9), 658.

Grunert, K. G., Sonntag, W. I., Glanz-Chanos, V. and Forum, S. 2018. Consumer interest in environmental impact, safety, health and animal welfare aspects of modern pig production: results of a cross-national choice experiment. *Meat Science* 137, 123-129.

Guy, J. H., Cain, P. J., Seddon, Y. M., Baxter, E. M. and Edwards, S. A. 2012. Economic evaluation of high welfare indoor farrowing systems for pigs. *Animal Welfare* 21(1) (Suppl.1), 19-24.

Hales, J., Moustsen, V. A., Nielsen, M. B. F. and Hansen, C. F. 2014. Higher preweaning mortality in free farrowing pens compared with farrowing crates in three commercial pig farms. *Animal* 8(1), 113-120.

Hales, J., Moustsen, V. A., Devreese, A. M., Nielsen, M. B. F. and Hansen, C. F. 2015. Comparable farrowing progress in confined and loose housed hyper-prolific sows. *Livestock Science* 171, 64-72.

Hamilton, W. D. 1964a. The genetical evolution of social behaviour. I. *Journal of Theoretical Biology* 7(1), 1-16.

Hamilton, W. D. 1964b. The genetical evolution of social behaviour. II. *Journal of Theoretical Biology* 7(1), 17-52.

Hansen, C. F., Hales, J., Weber, P. M., Edwards, S. A. and Moustsen, V. A. 2017. Confinement of sows 24 h before expected farrowing affects the performance of nest building behaviours but not progress of parturition. *Applied Animal Behaviour Science* 188, 1-8.

Hansen, L. U. 2018. Test of 10 different farrowing pens for loose-housed sows. Report no. 1803. SEGEs, Denmark.

Hartsock, T. G. and Barczewski, R. A. 1997. Prepartum behaviour in swine: effects of pen size. *Journal of Animal Science* 75(11), 2899-2904.

Hasan, S., Orro, T., Valros, A., Junnikkala, S., Peltoniemi, O. and Oliviero, C. 2019. Factors affecting sow colostrum yield and composition, and their impact on piglet growth and health. *Livestock Science* 227, 60-67.

Heidinger, B., Stinglmayr, J., Maschat, K., Oberer, M., Kuchling, S. and Baumgartner, J. 2018. Summary of the Austrian project "pro-SAU": evaluation of novel farrowing systems with possibility for the sow to move 2410:Eval. AFB Supplement zum Abschlussbericht des Projekts Pro-SAU.

Herpin, P., Damon, M. and Le Dividich, J. 2002. Development of thermoregulation and neonatal survival in pigs. *Livestock Production Science* 78(1), 25-45.

Hessel, E. F., Reiners, K. and Van den Weghe, H. F. A. 2006. Socializing piglets before weaning: effects on behavior of lactating sows, pre-and postweaning behavior, and performance of piglets. *Journal of Animal Science* 84(10), 2847-2855.

Homedes, J., Salichs, M., Sabaté, D., Sust, M. and Fabre, R. 2014. Effect of ketoprofen on pre-weaning piglet mortality on commercial farms. *The Veterinary Journal* 201(3), 435-437.

Honeyman, M. S., Roush, W. B. and Penner, A. D. 1998. Pig crushing mortality by hut type in outdoor farrowing. Annual Progress Report, Iowa State University, Ames, pp. 16-17.

Horrell, I. and Bennett, J. 1981. Disruption of teat preferences and retardation of growth following cross-fostering of 1-week-old pigs. *Animal Science* 33(1), 99-106.

Hötzel, M. J., Pinheiro Machado F°, L. C., Wolf, F. M. and Costa, O. A. D. 2004. Behaviour of sows and piglets reared in intensive outdoor or indoor systems. *Applied Animal Behaviour Science* 86(1-2), 27-39.

Hunt, K. and Petchey, A. M. 1987. A study of the environmental preferences of sows around farrowing. *Farm Building Progress* 89, 11-14.

Ison, S. H., Wood, C. M. and Baxter, E. M. 2015. Behaviour of pre-pubertal gilts and its relationship to farrowing behaviour in conventional farrowing crates and loose-housed pens. *Applied Animal Behaviour Science* 170, 26-33.

Ison, S. H., Clutton, R. E., Di Giminiani, P. and Rutherford, K. M. 2016. A review of pain assessment in pigs. *Frontiers in Veterinary Science* 3, 108.

Ison, S. H., Jarvis, S., Hall, S. A., Ashworth, C. J. and Rutherford, K. M. D. 2018. Periparturient behavior and physiology: further insight into the farrowing process for primiparous and multiparous sows. *Frontiers in Veterinary Science* 5, 122.

Jarrett, S. and Ashworth, C. J. 2018. The role of dietary fibre in pig production, with a particular emphasis on reproduction. *Journal of Animal Science and Biotechnology* 9(1), 1-11.

Jarvis, S., Lawrence, A. B., McLean, K. A., Deans, L. A., Chirnside, J. and Calvert, S. K. 1997. The effect of environment on behavioural activity, ACTH, b-endorphin and cortisol in pre-parturient gilts. *Animal Science* 65(3), 465-472.

Jarvis, S., Lawrence, A. B., McLean, K. A., Chirnside, J., Deans, L. A. and Calvert, S. K. 1998. The effect of environment on plasma cortisol and b-endorphin in the parturient pig and the involvement of endogenous opioids. *Animal Reproduction Science* 52(2), 139-151.

Jarvis, S., McLean, K. A., Calvert, S. K., Deans, L. A., Chirnside, J. and Lawrence, A. B. 1999. The responsiveness of sows to their piglets in relation to the length of parturition and the involvement of endogenous opioids. *Applied Animal Behaviour Science* 63(3), 195-207.

Jarvis, S., van der Vegt, B. J., Lawrence, A. B., McLean, K. A., Deans, L. A., Chirnside, J. and Calvert, S. K. 2001. The effect of parity and environmental restriction on behavioural and physiological responses of pre-parturient pigs. *Applied Animal Behaviour Science* 71(3), 203-216.

Jarvis, S., Calvert, S. K., Stevenson, J., van Leeuwen, N. and Lawrence, A. B. 2002. Pituitary-adrenal activation in pre-parturient pigs (Sus scrofa) is associated with behavioural restriction due to lack of space rather than nesting substrate. *Animal Welfare* 11, 371-384.

Jarvis, S., Reed, B. T., Lawrence, A. B., Calvert, S. K. and Stevenson, J. 2004. Peri-natal environmental effects on maternal behaviour, pituitary and adrenal activation, and the progress of parturition in the primiparous sow. *Animal Welfare* 13(2), 171-181.

Jarvis, S., D'Eath, R. B., Robson, S. K. and Lawrence, A. B. 2006. The effect of confinement during lactation on the hypothalamic-pituitary-adrenal axis and behaviour of primiparous sows. *Physiology and Behavior* 87(2), 345–352.

Jensen, P. 1986. Observations on the maternal behaviour of free-ranging domestic pigs. *Applied Animal Behaviour Science* 16(2), 131–142.

Jensen, P. 1993. Nest building in domestic sows: the role of external stimuli. *Animal Behaviour* 45(2), 351–358.

Jensen, P. 2002. Behaviour of pigs. In: P Jensen (Eds), *The Ethology of Domestic Animals*, pp. 159–172. CABI Publishing, Wallingford, UK.

Jensen, P. and Toates, F. M. 1993. Who needs 'behavioural needs'? Motivational aspects of the needs of animals. *Applied Animal Behaviour Science* 37(2), 161–181.

Jones, G. M., Edwards, S. A., Sinclair, A. G., Gebbie, F. E., Rooke, J. A., Jagger, S. and Hoste, S. 2002. The effect of maize starch or soya-bean oil as energy sources in lactation on sow and piglet performance in association with sow metabolic state around peak lactation. *Animal Science* 75(1), 57–66.

King, R. L., Baxter, E. M., Matheson, S. M. and Edwards, S. A. 2019. Temporary crate opening procedure affects immediate post-opening piglet mortality and sow behaviour. *Animal: An International Journal of Animal Bioscience* 13(1), 189–197.

Kobek-Kjeldager, C., Moustsen, V. A., Theil, P. K. and Pedersen, L. J. 2020. Effect of litter size, milk replacer and housing on production results of hyper-prolific sows. *Animal* 14(4), 824–833.

Kyriazakis, I. and Edwards, S. A. 1986. The effects of split suckling on behaviour and performance of piglets. *Applied Animal Behaviour Science* 16(1), 92.

Ladewig, J., Kloeppel, P. and Kallweit, E. 1984. A case of "reversed cannibalism" the piglets damaging the sow [vulvar lesions]. *Annales de Recherches Veterinaires (France)*. Annals of veterinary research, 15(2), pp.275–277.

Lawrence, A. B., McLean, K. A., Jarvis, S., Gilbert, C. L. and Petherick, J. C. 1997. Stress and parturition in the pig. *Reproduction in Domestic Animals* 32(5), 231–236.

Lawrence, A. B., Petherick, J. C., McLean, K. A., Deans, L. A., Chirnside, J., Gaughan, A., Clutton, E. and Terlouw, E. M. C. 1994. The effect of environment on behaviour, plasma cortisol and prolactin in parturient sows. *Applied Animal Behaviour Science* 39(3–4), 313–330.

Lawrence, A. B., Petherick, J. C., Mclean, K. and Gilbert, C. 1992. Mediation of stress-induced inhibition of oxytocin in farrowing sows by endogenous opioids. *Proceedings of the British Society of Animal Production* 1992, 56.

Leeb, B., Leeb, C., Troxler, J. and Schuh, M. 2001. Skin lesions and callosities in group-housed pregnant sows: animal-related welfare indicators. *Acta Agriculturae Scandinavica, Section A – Animal Science* 51(S30), 82–87.

Lensink, B. J., Ofner-Schröck, E., Ventorp, M., Zappavigna, P., Flaba, J., Georg, H. and Bizeray-Filoche, D. 2013. In Andres Aland, Thomas Banhazi (Eds) Lying and walking surfaces for cattle, pigs and poultry and their impact on health, behaviour and performance. In: *Livestock Housing: Modern Management to Ensure Optimal Health and Welfare of Farm Animals*, pp. 2509–2514. Wageningen Academic Publishers, Wageningen, The Netherlands.

Lewis, E., Boyle, L. A., Brophy, P., O'Doherty, J. V. and Lynch, P. B. 2005. The effect of two piglet teeth resection procedures on the welfare of sows in farrowing crates. Part 2. *Applied Animal Behaviour Science* 90(3–4), 251–264.

Lohmeier, R. Y., Grimberg-Henrici, C. G. E., Büttner, K., Burfeind, O. and Krieter, J. 2020. Farrowing pens used with and without short-term fixation impact on reproductive traits of sows. *Livestock Science* 231, 103889.

Loisel, F., Farmer, C., Ramaekers, P. and Quesnel, H. 2013. Effects of high fiber intake during late pregnancy on sow physiology, colostrum production, and piglet performance. *Journal of Animal Science* 91(11), 5269-5279.

Mainau, E., Ruiz-de-la-Torre, J. L., Dalmau, A., Salleras, J. M. and Manteca, X. 2012. Effects of meloxicam (Metacam®) on post-farrowing sow behaviour and piglet performance. *Animal: An International Journal of Animal Bioscience* 6(3), 494-501.

Mainau, E., Temple, D. and Manteca, X. 2016. Experimental study on the effect of oral meloxicam administration in sows on pre-weaning mortality and growth and immunoglobulin G transfer to piglets. *Preventive Veterinary Medicine* 126, 48-53.

Malmkvist, J., Pedersen, L. J., Kammersgaard, T. S. and Jørgensen, E. 2012. Influence of thermal environment on sows around farrowing and during the lactation period. *Journal of Animal Science* 90(9), 3186-3199.

Marchant, J. N., Broom, D. M. and Corning, S. 2001. The influence of sow behaviour on piglet mortality due to crushing in an open farrowing system. *Animal Science* 72(1), 19-28.

Martin, J. E., Ison, S. H. and Baxter, E. M. 2015. The influence of neonatal environment on piglet play behaviour and post-weaning social and cognitive development. *Applied Animal Behaviour Science* 163, 69-79.

Mattsson, B., Susic, Z. and Lundeheim, N. 2004. Time for care and comfort. In Proceedings of 18th IPVS Congress, Hamburg, Germany, p. 800.

Middelkoop, A., Choudhury, R., Gerrits, W. J. J., Kemp, B., Kleerebezem, M. and Bolhuis, J. E. 2018. Dietary diversity affects feeding behaviour of suckling piglets. *Applied Animal Behaviour Science* 205, 151-158.

Middelkoop, A., Costermans, N., Kemp, B. and Bolhuis, J. E. 2019. Feed intake of the sow and playful creep feeding of piglets influence piglet behaviour and performance before and after weaning. *Scientific Reports* 9(1), 16140.

Morgan, T., Pluske, J., Miller, D., Collins, T., Barnes, A. L., Wemelsfelder, F. and Fleming, P. A. 2014. Socialising piglets in lactation positively affects their post-weaning behaviour. *Applied Animal Behaviour Science* 158, 23-33.

Mount, L. E. 1967. The heat loss from new born pigs to the floor. *Research in Veterinary Science* 8(2), 175-186.

Mount, L. E. 1968. *The Climatic Physiology of the Pig*. Edward Arnold, London, UK.

Moustsen, V. A. and Poulsen, H. L. 2004. Anbefalinger vedr. dimensioner på fareboks og kassesti. *Landsudvalget for Svin, Danske Slagterier, Notat* nr. 414.

Moustsen, V. and Jensen, T. 2008. *Inventar til forbedring af hygiejne i stier til løsgående farende og diegievende søer*. Notat, Nr. 809, Dansk Svineproduktion.

Moustsen, V. A., Pedersen, L. J. and Jensen, T. 2007. Afprøvning af stikoncepter til løsgående farende og diegievende søer. Meddelelse (805): 1-25. Dansk Svineproduktion.

Moustsen, V. A., Lahrmann, H. P. and D'Eath, R. B. 2011. Relationship between size and age of modern hyper-prolific crossbred sows. *Livestock Science* 141(2-3), 272-275.

Moustsen, V. A., Hales, J., Lahrmann, H. P., Weber, P. M. and Hansen, C. F. 2013. Confinement of lactating sows in crates for 4 days after farrowing reduces piglet mortality. *Animal* 7(4), 648-654.

Mouttotou, N. and Green, L. E. 1999. Incidence of foot and skin lesions in nursing piglets and their association with behavioural activities. *Veterinary Record* 145(6), 160-165.

Muns, R., Silva, C., Manteca, X. and Gasa, J. 2014. Effect of cross-fostering and oral supplementation with colostrums on performance of newborn piglets. *Journal of Animal Science* 92(3), 1193-1199.

Muns, R., Manteca, X. and Gasa, J. 2015. Effect of different management techniques to enhance colostrum intake on piglets' growth and mortality. *Animal Welfare* 24(2), 185-192.

Muns, R., Malmkvist, J., Larsen, M. L. V., Sørensen, D. and Pedersen, L. J. 2016. High environmental temperature around farrowing induced heat stress in crated sows. *Journal of Animal Science* 94(1), 377-384.

Muns, R., Nuntapaitoon, M. and Tummaruk, P. 2017. Effect of oral supplementation with different energy boosters in newborn piglets on pre-weaning mortality, growth and serological levels of IGF-I and IgG. *Journal of Animal Science* 95(1), 353-360.

Nissen, S., Faidley, T. D., Zimmerman, D. R., Izard, R. and Fisher, C. T. 1994. Colostral milk fat percentage and pig performance are enhanced by feeding the leucine metabolite β-hydroxy-β-methyl butyrate to sows. *Journal of Animal Science* 72(9), 2331-2337.

Nowland, T. L., van Wettere, W. H. E. J. and Plush, K. J. 2019. Allowing sows to farrow unconfined has positive implications for sow and piglet welfare. *Applied Animal Behaviour Science* 221, 104872.

Ocepek, M. and Andersen, I. L. 2017. What makes a good mother? Maternal behavioural traits important for piglet survival. *Applied Animal Behaviour Science* 193, 29-36.

Ocepek, M., Goold, C. M., Busančić, M. and Aarnink, A. J. A. 2018. Drinker position influences the cleanness of the lying area of pigs in a welfare-friendly housing facility. *Applied Animal Behaviour Science* 198, 44-51.

Oliviero, C., Heinonen, M., Valros, A. and Peltoniemi, O. 2010. Environmental and sow-related factors affecting the duration of farrowing. *Animal Reproduction Science* 119(1-2), 85-91.

O'Mahony, S. M., Marchesi, J. R., Scully, P., Codling, C., Ceolho, A. M., Quigley, E. M., Cryan, J. F. and Dinan, T. G. 2009. Early life stress alters behavior, immunity, and microbiota in rats: implications for irritable bowel syndrome and psychiatric illnesses. *Biological Psychiatry* 65(3), 263-267.

Oostindjer, M., Bolhuis, J. E., Mendl, M., Held, S., Gerrits, W., Van den Brand, H. and Kemp, B. 2010. Effects of environmental enrichment and loose housing of lactating sows on piglet performance before and after weaning. *Journal of Animal Science* 88(11), 3554-3562.

Oostindjer, M., van den Brand, H., Kemp, B. and Bolhuis, J. E. 2011a. Effects of environmental enrichment and loose housing of lactating sows on piglet behaviour before and after weaning. *Applied Animal Behaviour Science* 134(1-2), 31-41.

Oostindjer, M., Bolhuis, J. E., Mendl, M., Held, S., van den Brand, H. and Kemp, B. 2011b. Learning how to eat like a pig: effectiveness of mechanisms for vertical social learning in piglets. *Animal Behaviour* 82(3), 503-511.

Oostindjer, M., Bolhuis, J. E., Simon, K., van den Brand, H. and Kemp, B. 2011c. Perinatal flavour learning and adaptation to being weaned: all the pig needs is smell. *PLoS ONE* 6(10), e25318.

Oostindjer, M., Kemp, B., van den Brand, H. and Bolhuis, J. E. 2014. Facilitating 'learning from mom how to eat like a pig' to improve welfare of piglets around weaning. *Applied Animal Behaviour Science* 160, 19-30.

Pajor, E. A. 1998. Parent-offspring conflict and its implications for maternal housing systems in domestic pigs. Doctoral dissertation, McGill University Libraries.

Pajor, E. A., Weary, D. M., Fraser, D. and Kramer, D. L. 1999. Alternative housing for sows and litters: 1. Effects of sow-controlled housing on responses to weaning. *Applied Animal Behaviour Science* 65(2), 105-121.

Patil, Y., Gooneratne, R. and Ju, X. H. 2020. Interactions between host and gut microbiota in domestic pigs: a review. *Gut Microbes*, 11(3), 310-334.

Pedersen, L. J., Jørgensen, E., Heiskanen, T. and Damm, B. I. 2006. Early piglet mortality in loose-housed sows related to sow and piglet behaviour and to the progress of parturition. *Applied Animal Behaviour Science* 96(3-4), 215-232.

Pedersen, L. J., Malmkvist, J. and Andersen, H. M. L. 2013. Housing of sows during farrowing: a review on pen design, welfare and productivity. In: Andres Aland and Thomas Banhazi (Eds), *Livestock Housing: Modern Management to Ensure Optimal Health and Welfare of Farm Animals,* pp. 285-297. Wageningen Academic Publishers. Wageningen, The Netherlands.

Pedersen, M. L., Moustsen, V. A., Nielsen, M. B. F. and Kristensen, A. R. 2011. Improved udder access prolongs duration of milk letdown and increases piglet weight gain. *Livestock Science* 140(1-3), 253-261.

Pedersen, T. F., Van Vliet, S., Bruun, T. S. and Theil, P. K. 2019. Feeding sows during the transition period—is a gestation diet, a simple transition diet, or a lactation diet the best choice? *Translational Animal Science* 4(1), 34-48.

Peng, X., Yan, C., Hu, L., Liu, Y., Xu, Q., Wang, R., Qin, L., Wu, C., Fang, Z., Lin, Y., Xu, S., Feng, B., Zhuo, Y., Li, J., Wu and Che, L. 2019. Effects of fat supplementation during gestation on reproductive performance, milk composition of sows and intestinal development of their offspring. *Animals* 9(4), 125.

Petherick, J. C. 1983. A biological basis for the design of space in livestock housing. In: Baxter, S. H., Baxter, M. R. and MacCormack, J. A. S. C. (Eds), *Farm Animal Housing and Welfare,* pp. 103-120. Martinus Nijoff Publisher, Boston, MA.

Phillips, P. A., Fraser, D. and Pawluczuk, B. 2000. Floor temperature preference of sows at farrowing. *Applied Animal Behaviour Science* 67(1-2), 59-65.

Plush, K., McKenny, L., Nowland, T., van Wettere, W. and Terry, R. 2019. *Reducing Sow Stress around Farrowing.* Report prepared for the Co-operative Research Centre for High Integrity Australian Pork. *(Pork CRC) project # 1C-114).*

Power, M. L. and Schulkin, J. 2013. Maternal regulation of offspring development in mammals is an ancient adaptation tied to lactation. *Applied and Translational Genomics* 2, 55-63.

Price, E. O., Hutson, G. D., Price, M. I. and Borgwardt, R. 1994. Fostering in swine as affected by age of offspring. *Journal of Animal Science* 72(7), 1697-1701.

Puppe, B. and Tuchscherer, A. 1999. Developmental and territorial aspects of suckling behaviour in the domestic pig (Sus scrofa f. domestica). *Journal of Zoology* 249(3), 307-313.

Quesnel, H., Farmer, C. and Devillers, N. 2012. Colostrum intake: influence on piglet performance and factors of variation. *Livestock Science* 146(2-3), 105-114.

Quiniou, N. and Noblet, J. 1999. Influence of high ambient temperatures on performance of multiparous lactating sows. *Journal of Animal Science* 77(8), 2124-2134.

Randall, J. M., Armsby, A. W. and Sharp, J. R. 1983. Cooling gradients across pens in a finishing piggery: II. *Journal of Agricultural Engineering Research* 28(3), 247-259.

Rantzer, D. and Svendsen, J. 2001. Slatted versus solid floors in the dung area of farrowing pens: effects on hygiene and pig performance, birth to weaning. *Acta Agriculturae Scandinavica, Section A - Animal Science* 51(3), 167-174.

Renaudeau, D. and Noblet, J. 2001. Effects of exposure to high ambient temperature and dietary protein level on sow milk production and performance of piglets. *Journal of Animal Science* 79(6), 1540-1548.

Rioja-Lang, F. C., Seddon, Y. M. and Brown, J. A. 2018. Shoulder lesions in sows: a review of their causes, prevention, and treatment. *Journal of Swine Health and Production* 26(2), 101-107.

Robert, S. and Martineau, G. P. 2001. Effects of repeated crossfosterings on preweaning behavior and growth performance of piglets and on maternal behavior of sows. *Journal of Animal Science* 79(1), 88-93.

Roehe, R. 1999. Genetic determination of individual birth weight and its association with sow productivity traits using Bayesian analyses. *Journal of Animal Science* 77(2), 330-343.

Rolandsdotter, E., Westin, R. and Algers, B. 2009. Maximum lying bout duration affects the occurrence of shoulder lesions in sows. *Acta Veterinaria Scandinavica* 51(1), 44.

Rooke, J. A., Sinclair, A. G., Edwards, S. A., Cordoba, R., Pkiyach, S., Penny, P. C., Penny, P., Finch, A. M. and Horgan, G. W. 2001. The effect of feeding salmon oil to sows throughout pregnancy on pre-weaning mortality of piglets. *Animal Science* 73(3), 489-500.

Rosvold, E. M. and Andersen, I. L. 2019. Straw vs. peat as nest-building material-the impact on farrowing duration and piglet mortality in loose-housed sows. *Livestock Science* 229, 203-209.

Rosvold, E. M., Kielland, C., Ocepek, M., Framstad, T., Fredriksen, B., Andersen-Ranberg, I., Næss, G. and Andersen, I. L. 2017. Management routines influencing piglet survival in loose-housed sow herds. *Livestock Science* 196, 1-6.

Rosvold, E. M., Newberry, R. C., Framstad, T. and Andersen, I. L. 2018. Nest-building behaviour and activity budgets of sows provided with different materials. *Applied Animal Behaviour Science* 200, 36-44.

Ryan, E. B., Fraser, D. and Weary, D. M. 2015. Public attitudes to housing systems for pregnant pigs. *PLoS ONE* 10(11), e0141878.

Schmitt, O., Baxter, E. M., Lawlor, P. G., Boyle, L. A. and O'Driscoll, K. 2019a. A single dose of fat-based energy supplement to light birth weight pigs shortly after birth does not increase their survival and growth. *Animals* 9(5), 227.

Schmitt, O., Baxter, E. M., Boyle, L. A. and O'Driscoll, K. 2019b. Nurse sow strategies in the domestic pig: I. Consequences for selected measures of sow welfare. *Animal* 13(3), 580-589.

Schmitt, O., Baxter, E. M., Boyle, L. A. and O'Driscoll, K. 2019c. Nurse sow strategies in the domestic pig: II. Consequences for piglet growth, suckling behaviour and sow nursing behaviour. *Animal* 13(3), 590-599.

Schmitt, O., O'Driscoll, K., Boyle, L. A. and Baxter, E. M. 2019d. Artificial rearing affects piglets pre-weaning behaviour, welfare and growth performance. *Applied Animal Behaviour Science* 210, 16-25.

Skok, J. and Škorjanc, D. 2014. Group suckling cohesion as a prelude to the formation of teat order in piglets. *Applied Animal Behaviour Science* 154, 15-21.

Sørensen, J. T. and Fraser, D. 2010. On-farm welfare assessment for regulatory purposes: issues and possible solutions. *Livestock Science* 131(1), 1-7.

Sørensen, J. T., Rousing, T., Kudahl, A. B., Hansted, H. J. and Pedersen, L. J. 2016. Do nurse sows and foster litters have impaired animal welfare? Results from a cross-sectional study in sow herds. *Animal* 10(4), 681–686.

Strathe, A. V., Bruun, T. S. and Hansen, C. F. 2017. Sows with high milk production had both a high feed intake and high body mobilization. *Animal* 11(11), 1913–1921.

Straw, B. E., Dewey, C. E. and Bürgi, E. J. 1998. Patterns of crossfostering and piglet mortality on commercial US and Canadian swine farms. *Preventive Veterinary Medicine* 33(1–4), 83–89.

Swan, K. M., Peltoniemi, O. A. T., Munsterhjelm, C. and Valros, A. 2018. Comparison of nest-building materials in farrowing crates. *Applied Animal Behaviour Science* 203, 1–10.

Telkänranta, H. and Edwards, S. A. 2018. Lifetime consequences of the early physical and social environment of piglets. In: Špinka, M. (Ed.), *Advances in Pig Welfare* pp. 101–136. Woodhead Publishing, Cambridge, UK.

Theil, P. K., Lauridsen, C. and Quesnel, H. 2014. Neonatal piglet survival: impact of sow nutrition around parturition on fetal glycogen deposition and production and composition of colostrum and transient milk. *Animal: An International Journal of Animal Bioscience* 8(7), 1021–1030.

Thodberg, K., Jensen, K. H., Herskin, M. S. and Jorgensen, E. 1999. Influence of environmental stimuli on nest building and farrowing behaviour in domestic sows. *Applied Animal Behaviour Science* 63(2), 131–144.

Thøgersen, J. and Zhou, Y. 2012. Chinese consumers' adoption of a "green" innovation—the case of organic food. *Journal of Marketing Management* 28(3–4), 313–333.

Thomsson, O., Sjunnesson, Y., Magnusson, U., Eliasson-Selling, L., Wallenbeck, A. and Bergqvist, A. S. 2016. Consequences for piglet performance of group housing lactating sows at one, two, or three weeks post-farrowing. *PLoS ONE* 11(6), e0156581.

Thorup, F., Wedel-Müller, R. L., Hansen, C. F., Kanitz, E. and Tuchscherer, M. 2015. Neonatal mortality in piglets is more due to lack of energy than lack of immunoglobulins. In: *International Conference on Pig Welfare: Improving Pig Welfare-What Are the Ways Forward?*, p. 84. Wageningen Academic Publishers; Copenhagen, Denmark

Tuchscherer, M., Puppe, B., Tuchscherer, A. and Tiemann, U. 2000. Early identification of neonates at risk: traits of newborn piglets with respect to survival. *Theriogenology* 54(3), 371–388.

Turner, S. P., Camerlink, I., Baxter, E. M., D'Eath, R. B., Desire, S. and Roehe, R. 2018. Breeding for pig welfare: opportunities and challenges. In: Špinka, M. (Ed.), *Advances in Pig Welfare*, pp. 399–414. Woodhead Publishing, Cambridge, UK.

Tybirk, P., Sloth, N. M. and Jorgensen, L. 2012. *Danish Nutrient Requirement Standards (in Danish: Normer for Naringsstoffer)* (17th rev. edn.). SEGES Pig Research Centre, Axelborg, Denmark.

Tybirk, P., Sloth, N. M., Sonderby, T. B.. and Kjeldsen, N. 2015. *Danish Nutrient Requirement Standards (in Danish Normer for Naringsstoffer)* (22th rev. edn.). SEGES Pig Research Centre, Axelborg, Denmark.

van Beirendonck, S., Schroijen, B., Bulens, A., Van Thielen, J. and Driessen, B. 2015. A solution for high production numbers in farrowing units? In: *International Conference on Pig Welfare: Improving Pig Welfare-What Are the Ways Forward*, pp. 85–85. Wageningen Academic Publishers, Copenhagen, Denmark.

van de Weerd, H. and Ison, S. 2019. Providing effective environmental enrichment to pigs: how far have we come? *Animals* 9(5), 254.

van Dixhoorn, I. D., Reimert, I., Middelkoop, J., Bolhuis, J. E., Wisselink, H. J., Koerkamp, P. W. G., Kemp, B. and Stockhofe-Zurwieden, N. 2016. Enriched housing reduces disease susceptibility to co-infection with porcine reproductive and respiratory virus (PRRSV) and Actinobacillus pleuropneumoniae (A. pleuropneumoniae) in young pigs. *PLoS ONE* 11(9), e0161832.

van Nieuwamerongen, S. E., Bolhuis, J. E., Van der Peet-Schwering, C. M. C. and Soede, N. M. 2014. A review of sow and piglet behaviour and performance in group housing systems for lactating sows. *Animal* 8(3), 448-460.

van Nieuwamerongen, S. E., Soede, N. M., van der Peet-Schwering, C. M. C., Kemp, B. and Bolhuis, J. E. 2015. Development of piglets raised in a new multi-litter housing system vs. conventional single-litter housing until 9 weeks of age. *Journal of Animal Science* 93(11), 5442-5454.

Verbeke, W. 2009. Stakeholder, citizen and consumer interests in farm animal welfare. *Animal Welfare* 18(4), 325-333.

Verdon, M., Morrison, R. S. and Rault, J. L. 2019. Sow and piglet behaviour in group lactation housing from 7 or 14 days post-partum. *Applied Animal Behaviour Science* 214, 25-33.

Wathes, C. and Whittemore, C. T. 2006. Environmental management of pigs. In: Kyriazakis, I. and Whittemore, C. T. (Eds), *Whittemore's Science and Practice of Pig Production*, pp. 533-592. Blackwell Publishing, Oxford, UK.

Weary, D. M., Pajor, E. A., Bonenfant, M., Fraser, D. and Kramer, D. L. 2002. Alternative housing for sows and litters. Part 4. *Applied Animal Behaviour Science* 76(4), 279-290.

Weary, D. M., Jasper, J. and Hötzel, M. J. 2008. Understanding weaning distress. *Applied Animal Behaviour Science* 110(1-2), 24-41.

Westin, R., Holmgren, N., Hultgren, J. and Algers, B. 2014. Large quantities of straw at farrowing prevents bruising and increases weight gain in piglets. *Preventive Veterinary Medicine* 115(3-4), 181-190.

Widowski, T. M. and Curtis, S. E. 1990. The influence of straw, cloth tassel, or both on the prepartum behaviour of sows. *Applied Animal Behaviour Science* 27(1-2), 53-71.

Wiegand, R. M., Gonyou, H. W. and Curtis, S. E. 1994. Pen shape and size: effects on pig behavior and performance. *Applied Animal Behaviour Science* 39(1), 49-61.

Williams, A. M., Safranski, T. J., Spiers, D. E., Eichen, P. A., Coate, E. A. and Lucy, M. C. 2013. Effects of a controlled heat stress during late gestation, lactation, and after weaning on thermoregulation, metabolism, and reproduction of primiparous sows. *Journal of Animal Science* 91(6), 2700-2714.

Wischner, D., Kemper, N. and Krieter, J. 2009. Nest-building behaviour in sows and consequences for pig husbandry. *Livestock Science* 124(1-3), 1-8.

Wu, G., Bazer, F. W., Johnson, G. A., Knabe, D. A., Burghardt, R. C., Spencer, T. E., Li, X. L. and Wang, J. J. 2011. Triennial Growth Symposium: important roles for L-glutamine in swine nutrition and production *Journal of Animal Science* 89(7), 2017-2030.

You, X., Li, Y., Zhang, M., Yan, H. and Zhao, R. 2014. A survey of Chinese citizens' perceptions on farm animal welfare. *PLoS ONE* 9(10), e109177.

Yun, J., Swan, K. M., Vienola, K., Kim, Y. Y., Oliviero, C., Peltoniemi, O. A. T. and Valros, A. 2014. Farrowing environment has an impact on sow metabolic status and piglet colostrum intake in early lactation. *Livestock Science* 163, 120-125.

Yun, J. and Valros, A. 2015. Benefits of prepartum nest-building behaviour on parturition and lactation in sows—a review. *Asian-Australasian Journal of Animal Sciences* 28(11), 1519-1524.

Yunes, M. C., von Keyserlingk, M. A. G. and Hötzel, M. J. 2017. Brazilian citizens' opinions and attitudes about farm animal production systems. *Animals* 7(10), 75.

Zanella, A. J. and Zanella, E. L. 1993. Nesting material used by free-range sows in Brazil. In: Nichelmann, M., Wierenga, H. K. and Braun, S. (Eds), *Proceedings of the 3rd Joint Meeting of the International Congress on Applied Ethology*, p. 411. Humboldt-Universitaet, Berlin.

Zoric, M., Nilsson, E., Lundeheim, N. and Wallgren, P. 2009. Incidence of lameness and abrasions in piglets in identical farrowing pens with four different types of floor. *Acta Veterinaria Scandinavica* 51(1), 23.

Zurbrigg, K. 2006. Sow shoulder lesions: risk factors and treatment effects on an Ontario farm. *Journal of Animal Science* 84(9), 2509-2514.

Chapter 2

Optimising pig welfare at the weaning and nursery stage

Nicole Kemper, University of Veterinary Medicine Hannover, Germany

1   Introduction
2   Minimising the impact of weaning
3   Optimised management: the human key factor
4   Conclusion
5   Future trends in research
6   Where to look for further information
7   References

1 Introduction

1.1 Background

The most frequently mentioned key production parameters in pig farming are litter size and daily weight gain in the fattening period. However, between farrowing and fattening, the rearing of the weaned piglets in the nursery unit represents a very important and delicate period in the life of a pig with a number of risks for their welfare. Also at this bridging stage, optimal care and profound knowledge about pigs are essential for enabling both pig health and welfare in the young animals and laying the foundation for successful fattening. Since weaning itself probably represents the most stressful event in the life of a pig, maintaining their welfare is of enormous importance in this period to prevent stress and serious drawbacks, such as weight check or diarrhoea. Weaning leads to a stress response with increased levels of corticotrophin-releasing factor (CRF) and cortisol after activation of the hypothalamic pituitary adrenal axis (Moeser et al., 2017). Together with activation of the immune system and compromised digestive and absorptive capacities of the gut due to physiological changes in structure and function, a reduction in feed efficiency and lean tissue deposition can be observed (Campbell et al., 2013; Spurlock, 1997). Weight losses of approximately 100–250 g on the first day post-weaning are caused by this stress, and undernutrition can be induced by the change from milk to solid feed (Hennig-Pauka et al., 2019; Le Dividich and Sève, 2000).

http://dx.doi.org/10.19103/AS.2020.0081.05

This loss is usually recovered by about four days post-weaning if no further adverse effects occur (Le Dividich and Sève, 2000).

The following statements refer mainly to commercial production systems with separate farrowing housings, after which weaned piglets are transported to another barn or compartment, the nursery unit. The particular challenge in the nursery stage is the provision of a good start to the weaners from the beginning. In contrast to weaning in wild boars or pigs in a (semi-) natural environment, which happens progressively until the 9th and 16th weeks of life (Jensen, 1986; Newberry and Wood-Gush, 1985), weaning in pig production is an abrupt procedure, usually occurring at an age around four weeks of life or less with several extreme changes. Over the last several decades, the weaning age has decreased significantly, as described for French piglets from about eight weeks in the 1950s to about 25 days today (Prunier et al., 2010). In the European Union, legal regulations demand a weaning age of at least 28 days of age with an exception for piglets that are moved into specialised compartments (emptied, thoroughly cleaned and disinfected, and separated from sow housings), which can be weaned seven days earlier to ease all-in-all-out batch management (Council Directive 2008/120/EC, 2008). In organic systems, piglets are allowed to suckle longer until weaning starts between six and eight weeks of age. The actual weaning age in commercial production is a compromise between producing a high number of piglets per sow and per year via short intervals between successive farrowing on one hand, versus having piglets robust enough to cope with the weaning stress on the other. While it has been shown that three to four weeks are the optimal weaning time to maximise sow performance (Levis, 1997), the weaned piglets are still immature at this age and very sensitive to weaning stress. Weaning stress is caused by abrupt changes in the piglets' life, including, among others, separation from the mother sow, dietary changes, exposure to a new environment and mixing with unacquainted pen mates. It can impair piglet welfare seriously if no appropriate counter or mitigation measures are applied. Weaned piglets in commercial conditions face concurrent stressors, especially directly after weaning. Each single stressor that can be minimised is helpful in reducing negative impacts on piglet welfare. The most obvious effect of decreased welfare in weaners, especially in barren environments, is the occurrence of abnormal behaviours; with no or only few physical- and mental-stimulating activities and too little space allowance, stereotypes or injurious behaviours such as tail- or ear-biting can occur (Beattie et al., 1995; Stafford, 2010). The etiology of these behaviours is often multifactorial; thus the provision of an adequate environment and the reduction of possible stressors are even more important.

The process of weaning and the following period demand high adaptive capabilities to help the piglets cope with stress. Even if the first critical days in the nursery stage are overcome, the stockperson has to have a careful eye on the growing pigs to ensure they become healthy fatteners without any declines in health and welfare. In general, several factors are attributed to pig welfare and have to be addressed carefully in the nursery unit: (1) space allowance; (2) structuring the housing environment in functional areas; (3) availability of food, water and rooting material; (4) social contacts; (5) cleanliness; (6) efficient regulation of the barn climate; and (7) management in order to enable a stable social hierarchy (Bracke et al., 2002a,b; Scipioni et al., 2009). The more elements from the farrowing unit can be offered in a similar way in the nursery, the smoother the abrupt post-weaning changes will be resolved.

1.2 The importance of enabling optimised welfare in nursery pigs and the key challenges

Welfare is a basic precondition for healthy pigs showing a high performance, most apparent in good growth and feed efficiency, at the nursery stage. Even more important, the opportunity to manage pig welfare is required by law in many countries and is an indicator for professional pig farming following the best practices. In the nursery, the foundation is laid for successful fattening, and, under commercial aspects, the aims are to maximise growth performance and reduce morbidity and mortality. From a welfare point of view, the weaning process with all of the accompanying changes in the lives of piglets combined with their very young age represents a potential tremendous strain. Welfare impairments can be classified as environmental, nutritional and social stress (Lallès et al., 2004), and subsequent health problems (Fig. 1). Keeping these negative impacts as minimal as possible is urgently needed because otherwise the piglets are more prone to diseases, grow less and are difficult to manage, especially when lagging behind the rest of the group. These shortcomings in the nursery stage usually cannot be made up later on in the piglets' life; bodyweight at weaning, and weight gain directly post-weaning can affect growth until the end of the fattening period (Pajor et al., 1991).

A major challenge is the sensitivity of the young piglets at weaning and in the first weeks after that. Weaners are vulnerable and immature piglets, and require far more care than finishing pigs. These piglets are fragile and inexperienced. Before weaning, the sow's milk is the single nutrient and energy source for the suckling piglets (Bøe, 1991). The separation from their mothers leaves them without a milk supply, warmth and motherly care, including touching, comforting and pheromones (Niekamp et al., 2007). Because of this separation and transport to an unknown environment, they become mentally and physically stressed, even before arriving at the nursery barn. The nursery barn represents a new physical and microbial environment. Until then, the farrowing pen was the only environment they knew. The new surroundings, together with the changes in barn climate and microclimate, represent

Social stressors
- separation from mother sow
- regrouping
- fighting
- mutilations by penmates

⇩

Nutritional stressors
- change to solid feed and water
- unfamiliar feed
- unfamiliar technical devices
- gastrointestinal changes
- fastening/starving

⇨

Environmental stressors
- transport
- unfamiliar environment
- low temperatures
- poor sanitary conditions
- crowding
- pathogen pressure
- improper handling

⇦

Piglet at weaning

⇩

Consequences
- growth check
- impaired welfare
- health problems

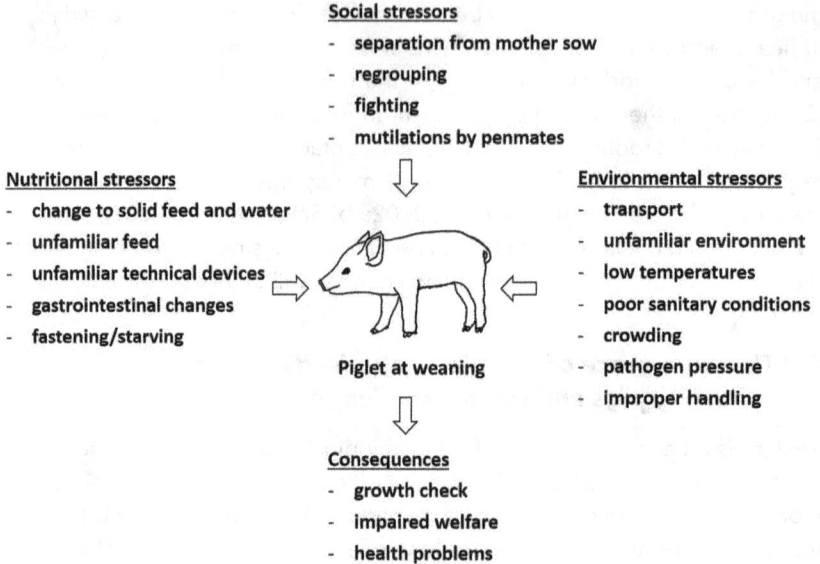

Figure 1 Potential stressors of different origins and possible consequences for piglets at weaning.

environmental stress. Nutritional stress results from the abrupt change from milk to solid feed, which goes along with a decreased nutrient intake and as a consequence, lower body heat. In turn, this decrease exacerbates environmental stress, especially in a very cold environment. Social stress is caused both by the separation from their mother and mixing with unknown piglets from other litters that can result in hierarchical fights. Moreover, weaning and movement to the nursery unit also represent a challenge to piglet health. Because health is one prerequisite to good welfare, it is of the utmost importance to prevent infections. Infections such as diarrhoea or respiratory diseases at weaning age are caused by pathogens present in the environment after insufficient cleaning and disinfection. These two factors are facilitated by not yet been fully developed immune response in the young animals and their lack of effective antibody formation. The immune system of weaners has not matured at this point, and even worse, the usual period of weaning falls within a critical lack of protective immunoglobulins in weaner blood. Their immature immune system is not yet capable of producing specific immunoglobulins (antibodies), and the immunoglobulins acquired via colostrum from their mother (maternal immunity) are already declining (Moeser et al., 2017). This lack of immunity appears between the second and fourth week of life, which is the exact time of weaning in commercial pig production systems. Negative effects, such as infectious diseases, especially post-weaning diarrhoea and even deaths, have been well described (Lallès et al., 2004; Weary et al., 2008). These negative effects can be aggravated when piglets are weaned with low weaning weights as could happen in large, uneven litters with a certain percentage of low-weighing and sometimes underdeveloped piglets (Hales et al., 2013; Koketsu et al., 2017). Only a very low percentage of piglets weighing less than 1 kg at birth were still alive at weaning in a study by Quiniou et al. (2002). In accordance, in another study, 86% of piglets weighing less than 0.80 kg at birth did not survive to weaning (Gondret et al., 2005). Birth weight and weaning weight correlate with one another, and as stated by Smith et al. (2007): 'Pigs that are heavier at birth are worth more money because they are likely to be heavier at the end of the nursery phase.' Therefore, special attention should be given to low-weight piglets from litters of highly prolific sows, not only in the farrowing unit (as reviewed by De Vos et al., 2014), but also in the subsequent weaning steps and while in the nursery stage.

2 Minimising the impact of weaning

2.1 The weaning process: preparation of procedures and facilities

Piglets are usually weaned at a certain age in order to empty the farrowing compartment and make room for the next farrowing batch. However, weaning is not only based on pure planning data, but also on the actual appearance of the piglet and its weight. Low birthweight piglets are defined in two ways; first as newborns with a birthweight below the 10th percentile of the mean birthweight of the litter, and second, with birthweight less than the mean birthweight minus two times the standard deviation (Cooper, 1975). Similarly, low-weaning weight piglets can be defined with the mean weaning weight of the batch as reference. Those piglets are at special risk for post-weaning welfare and health impairments.

The weaned piglets should be uniform and at a comparable weight level. All weaning preparations for sows and to where to move the single sow (slaughter or service) and the weaned piglets have to be completed in due time. The impact of weaning can be reduced by a careful planning of all processes and a thorough preparation of the facilities. The preparation of the nursery facilities in such a way that the piglets' environmental needs are adequately reflected is essential for giving them a good start in life. Good preparation includes calculations of the space and respective number of pens needed for the whole batch of weaned piglets. An adequate space allowance is important for weaner pigs, taking into account their expected weights when leaving the nursery unit for the calculation of stocking density. Legal requirements have to be met even though it has been shown that they often fail to provide a high amount of free space for different species-specific behaviours (Fels et al., 2019).

At the time of weaning in most production systems, the piglets are moved from the farrowing barns into the nursery unit. Only on some farms, the sows are removed from the farrowing pen, and the piglets stay there until they are brought to the fattening units. The following details refer mainly to the first situation. In general, experienced and trained farm personnel form an indispensable prerequisite for a smooth weaning process by keeping the negative impacts on the piglets as low as possible. When chasing piglets, a driving board should be used and all movements should be performed in calm manner.

Another point to consider in organising the weaning process is animal transportation. In the United States, Canada and other countries, transportation of weaned piglets to separate production facilities is a common practice. Transportation duration should be as short as possible, however, for the United States, duration up to 24 h in trailers without supplemental heat, feed and water are described (Lewis and Berry, 2006; Sutherland et al., 2009). Each transport affects the welfare of the newly weaned piglets as the animals experience stress, as reviewed comprehensively by Sutherland et al. (2014). The extent of stress and impaired welfare, especially by the concurrent weaning and transport, is influenced by season, respectively temperature, space allowance and duration of the transport and further aspects such as mixing with unfamiliar conspecifics. In addition, handling at loading, vibrations, noise, acceleration, water and feed withdrawal contribute to the extent of stress (Sutherland et al., 2014). In Northern America, segregated early weaned piglets are typically 14-21 days of age (Lewis and Berry, 2006), while in European countries and other regions of the world, they are usually older when weaned and transported, and therefore more robust than at younger age. Concerning mortality in weaners at transport, only limited data is available. Averós et al. (2010) stated that in 13.8% of 109 European transports, including 58 682 piglets, deaths occurred, and 0.07% of all transported piglets were found dead on arrival. However, these data did not refer to freshly weaned piglets, but to animals with 85-100 days of age, which were weaned between 21 and 28 days of age (Averós et al., 2010). Nevertheless, the identified risk factors, duration of journey and mean outside temperature with significant effects on mortality are most probably also true for freshly weaned piglets, as also shown in transport simulations by Berry and Lewis (2001). An overview of the special situation in transport of early weaned piglets is given by Lewis (2008), stressing the importance of ambient temperature and duration as well. Regarding ambient temperature, both summer (Lewis and Berry, 2006) and winter (Lewis and Berry, 2006; Wamnes et al., 2006) were described to negatively affect weaners' growth check and recovery time. Besides ambient conditions, the genetic background of the piglets might also influence the actual effect of transport on welfare, as stated by Averós et al. (2009). Sutherland et al. (2014) recommended to provide 0.06

or 0.07 m²/pig at short (< 1 h) transportation trips in summer, and concluded that transports of up to 6 h might not cause more stress than weaning alone, but further studies on the short- and long-term effects of transport on weaners' welfare are needed to develop detailed recommendations. Each transport has an effect on piglet welfare, and therefore, should be kept as short as possible. Transportation vehicles, whether motorised or simple trolleys, should be equipped with adequate bedding to avoid physical damages to or cooling of the piglets.

In order to minimise the negative impact of weaning on pig welfare from environmental changes, the nursery barn must be prepared carefully before the freshly weaned piglets arrive. These preparations include measures to ensure both proper health and welfare by minimising infection risks and providing an adequate environment by meeting housing requirements of young piglets. Housing requirements of weaners are summarized also by Madec et al. (2003).

Independent of the housing system and the exact design, such as flooring or bedding, the facilities must be thoroughly cleaned and disinfected before the piglets arrive. The cleaning is crucial to remove all possible pathogenic microorganisms originating from the preceding batch of nursery piglets. Proper hygiene management based on all-in/all-out management and cleaning and disinfection of all housing facilities has been shown to be of the utmost importance for the pigs' health (Madec et al., 1998). As shown by Le Floch et al. (2006), piglets weaned with 28 days of life and housed in a housing facility of high sanitary status showed significantly higher final body weight 45 days after weaning compared to piglets kept in poor sanitary conditions; however, they also received antibiotic supplementation via standard diet and were kept in single cages. Sanitation measures are often unpopular tasks, but the importance of their careful performance cannot be stressed enough. There are many possible pitfalls in cleaning and disinfection, but the most important points to consider in the nursery are listed:

- thoroughly remove faecal soiling from the previous batch of piglets by soaking and cleaning as soon as the previous group of pigs is moved to the fattening unit;
- dust should be removed, including from equipment that is not easy to reach, such as fans, air ducts, and feed and water pipes above head-height, because pathogens attach to dust and can survive and stay infectious over extremely long periods of time (Schulz et al., 2016);
- use effective disinfectants with a broad spectrum, and, if needed, also specifically against known, identified pathogens;
- avoid gaps in effectiveness of disinfection caused by protein failures, soap failures or low temperature failures; and

- also clean and disinfect technical equipment, such as trolleys on which the piglets are held or moved.

These cleaning and disinfection procedures should be incorporated into a general farm biosecurity concept, including, among others, proper batch management with all-in/all-out, boot hygiene, shower facilities at the entrance, and rodent control. One important point to consider after cleaning and disinfection is the drying of the barn. Drying is essential for a proper disinfection. After cleaning and before application of the disinfectant drying avoids dilution effects of the disinfectant by puddles, and after disinfection drying prevents the animals from drinking and ingesting disinfectant residues. No wet spots should be present anymore when the piglets arrive.

Even with the most careful sanitising, exposures of the pigs to environmental pathogens cannot be avoided. To a certain degree, this exposure is even desirable, because it can lead to a resilient immunity when the immune system of the pigs is fully developed. However, due to the lack of immune competence in weaners at the age of two-four weeks, this option is not applicable in the early nursery stage, and instead, everything has to be done to ensure the piglets are strong enough to cope with all stressors that occur immediately post-weaning. Besides the reduction in pathogen concentrations and possible transmission routes in the weaners' environment, the immediate surroundings should be prepared and equipped in such a way that help the pigs to adjust and, at a minimum, does not represent a source of further health and welfare risks.

First, the temperature of the pens is the key; piglets at weaning experience cold stress because they are removed from the farrowing environment providing a heated creep area for them as well as from their mothers, which also provided warmth and other nurturing functions. Piglets lack thermoregulation capacities, meaning that they cannot regulate their body temperature sufficiently because they do not possess the necessary insulating fat reserves at this point and additionally, cannot consume satisfactory amounts of feed as an energy source in order to maintain their body temperatures high enough, especially after the sudden change from milk to solid feed. The temperature in the pen at the animals' body height in the first week after weaning should be between 28°C and 32°C for piglets weighing about 5 kg; especially in the lying area this temperature is required (Mavromichalis, 2001; Muirhead et al., 2013). For weaners, the lower critical temperature ranges from 26°C to 28°C during the first week and from 23°C to 24°C during the second week after weaning (Noblet et al., 2001). The nursery unit must be heated up and prewarmed to these temperatures before the weaners arrive. Keeping this temperature at the animals' body height depends significantly on the design and types of equipment in the pen. In straw-based systems, a sufficient amount of bedding material should be provided to keep the piglets warm even without an additional heat source.

Floor design is of special importance; in slatted floor systems, drafts can develop easily, and depending on the floor type, ammonia emission rates can increase (Aarnink et al., 1997). Lying mats (made of rubber or plywood), pen divisions, and shelters or covers can help to establish segments with an optimal microclimate. In this way, structuring the pen into separate dunging and lying areas is supported. Regarding flooring, a detailed overview was provided by the EFSA Panel on Animal Health and Welfare (2005). The floor has to be designed in such a way that the risk of claw and leg injuries due to inappropriate slat and gap widths and rough materials is minimized. In a British study, KilBride et al. (2009) compared the impact of different floor types on the occurrence of foot and limb lesions, and showed that the presence of bursitis, capped hock and calluses was lowest in rearing pigs between 6 and 22 weeks of age housed on soil (outdoor rearing), followed by solid concrete with a mixture of deep and sparse bedding, and highest in pigs housed on fully slatted floors. The positive effect of bedding was confirmed by Munsterhjelm et al. (2009).

Concerning ventilation for freshly weaned pigs, an air renewal rate of not more than 0.1 m^2/h/kg of body weight is recommended (Mavromichalis, 2001). As appropriate for pigs at all ages, relative air humidity should range between 60% and 80%. In a two-year-experiment, Done et al. (2005) found no significant effects of ammonia levels of up to 37 ppm and dust concentrations of up to 9.9 mg/m^3 (inhalable fraction) on general pathology, respiratory tract pathology and microbiological findings in lungs, tracheae and nasal cavities in weaners kept under these conditions for 37 days. The authors conclude that such high concentrations are unlikely to be associated with respiratory diseases in weaners under the conditions of a good all-in-all-out management with thorough cleaning and disinfection, a high standard of nutrition and husbandry and a special attention to pigs' welfare (Done et al., 2005). However, as weaned piglets show an aversion against ammonia at levels of and above 10 ppm (Jones et al., 1996), ammonia concentrations in the nursery should be kept below this concentration to promote welfare.

Before taking the weaners to the new facility, the equipment, especially feeders and drinking nipples or other sources of water must be checked and be in a proper, functioning condition. The piglets must acclimate to the water source, especially when a different type of water cup or nipple was provided in the farrowing unit. Sufficient amounts of fresh water, without any deficiencies in quality should be accessible ad libitum, with recommended water flow rates at nipple drinkers of 500 mL/min (Scipioni et al., 2009). Besides the drinkers' efficiency, the number and the position of any watering unit are highly relevant. Before the piglets arrive, fresh, clean water should be poured into cups or should trickle from the nipples to help them to become familiar with these devices. All other equipment, such as manipulable materials, should be clean at arrival. Bedding or enrichment material should be dry and located in the assigned areas and at room temperature. In addition, the hospital pens should also be prepared before arrival.

2.2 Environmental stress: strategies to reduce possible welfare impacts

For the provision of a suitable environment, not only the technical set-up has to be animal-friendly, but also the climate is of key importance. As already mentioned, focus should be laid on the optimal barn temperature for the respective age of the piglets. In the first two weeks post-weaning, piglets need, in addition to the above-mentioned temperatures at animals' body height, a room temperature of about 25–30°C in order to maintain thermal neutrality, depending on their weight and the floor type (Muirhead et al., 2013). Even though it was shown in experiments that piglets weaned at three weeks of age can adapt to long-term cold exposures (12°C for three weeks), the physiological consequences as changes in glucose turnover, thyroid hormones and catecholamines have significant negative impacts especially on weight gain, but possibly also on health (Herpin et al., 1987). Le Dividich and Herpin (1994) summarized the effects of climatic conditions on performance, metabolism and health of weaned pigs in a review. With increasing weight and in bedded systems, temperatures can be lower. In preference experiments, weaners aged between four and eight weeks preferred straw-bedded areas for resting at ambient temperatures of 18–21°C, while at 25–27°C, concrete floors were selected (Fraser, 1985). This indicates that the weaners' preference for type of flooring depends strongly on temperature. Instead of only checking and comparing the detailed measured values of ambient climate via different climate data loggers, the animals and their behaviour should be observed at each tour through the compartments. Their behaviour is the ultimate indicator of climatic comfort and welfare in general. Healthy pigs in their climatic comfort zone do not huddle together or spread out as much as possible. They do not pant. It has to be stressed that pigs in conventional nurseries without any outdoor access are kept in a manmade environment for 24 h a day. Their possibilities to choose or avoid unfavourable environmental conditions are very limited. For instance, if the underfloor heating is too hot, piglets will avoid it as far as possible, but with that avoidance, available space for free movements is limited (Fig. 2). The responsibility for the provision of an adequate animal-friendly environment lies with the farmer and the stockmen on the farm.

Pen cleanliness has to be checked and controlled daily, and bedding and enrichment material has to be renewed if necessary. From a technical point of view, all devices, such as fans, heaters, lights, feeders and drinkers have to be

(a)

(b)

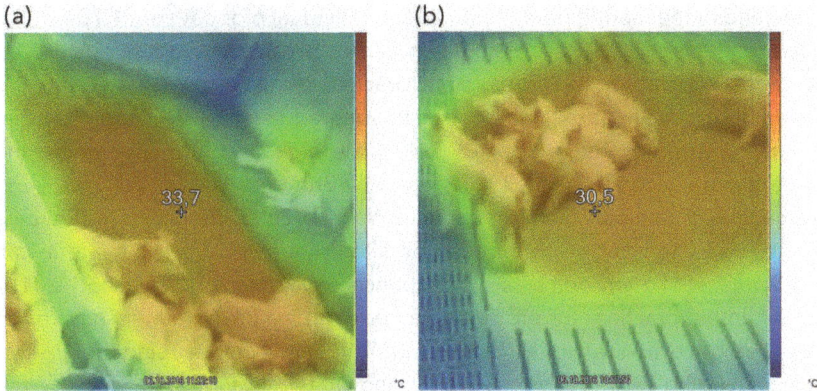

Figure 2 Piglets avoid the over-heated heating panel (a) and use the correctly heated one (b) (Photo with thermal imaging camera: ITTN, TiHo).

checked daily for proper function. Even with careful preparation of the facilities, movement into a new, unknown environment is a stressful situation for the pigs. In order to avoid health and welfare impairments, not only a good preparation of the pure infrastructure is necessary, but the pens should also be large enough and equipped with enrichment to smooth negative impacts caused by other stressors, especially social stress. If the space is limited, high animal density and crowding impacts animal welfare and animal health, and with those negative effects, also influences performance. Several studies showed the negative effect of decreased space allowance on feed intake and weight gain in weaners (Brumm et al., 2001; Stojanac et al., 2014). Besides significantly lower daily weight gains and higher feed conversion rates, Stojanac et al. (2014) detected a significantly increased occurrence of *Salmonella* species, higher mortalities and higher culling rates in weaners kept at the highest stocking rate of 0.16 m² available space per pig, compared to 0.25 m² and 0.32 m² per pig. These findings emphasise the importance of adequate space allowance not only for production parameters, but also for the safety of pig meat for human nutrition. In a South Korean study, reduced space allowance affected not only daily weight gain negatively, but resulted in a linear increase in plasma-cortisol and Interleukin-1ß in weaners (Oh et al., 2010). From their results, the authors suggest a space allocation between 0.30 m² and 0.43 m² per animal for weaners until 15 kg of bodyweight to maximize growth performance and immune response (Oh et al., 2010). In the study by Smith et al. (2004), increasing space allowance from 0.23 m² per pig to 0.35 m² was associated with increased body weight at approximately ten weeks of age. Regarding covered floor spaces by freshly weaned piglets, an individual piglet of 7 kg bodyweight covered 793 cm² in standing position, 761 cm² in sitting posture and 838 cm² in a recumbent position (Fels et al., 2018a). Covered space increased linearly with weight, and at the end of the rearing period with 25 kg, individual pigs covered 1.704 cm² in a standing position, 1.687 cm² sitting and 1.798 cm² in a recumbent posture, on average (Fels et al., 2018a). These static space data can be useful in determining the amount of free space, in relation to the offered pen size, which remains per pig after deducting the space needed by the pig's body. In more detail, the influence of space allowance is discussed by Brumm and Gonyou (2001) and by the EFSA Panel on Animal Health and Welfare (2005). Especially, when pigs are crowded, the gut microflora can be altered and pathogenic bacteria can proliferate, with diarrhoea as a consequence (Jayaraman and Nyachoti, 2017). Management practices for promoting a good gut health in weaners have been reviewed by Jayaraman and Nyachoti (2017) in detail, emphasising the need to avoid crowding stress in addition to feeding strategies and sanitation.

Environmental enrichment after weaning has been shown to improve feed efficiency and reduce the incidence of diarrhoea (Oostindjer et al., 2010). Compared with a barren environment, moderate bedding has positive impacts not only on welfare and in the reduction of post-weaning diarrhoea, but also in increasing growth of the weaners (Munsterhjelm et al., 2009). Increased space allowance alone does not result in decreased aggression or other harmful behaviour, but combined with enrichment, positive effects were shown (Beattie et al., 1996). Martin et al. (2015) examined the influence of an enriched environment in the farrowing unit on play behaviour and on post-weaning social and cognitive development in weaners, and concluded that a constant enrichment is needed to regularly stimulate play behaviour in order to facilitate animal welfare. As it is highly beneficial to provide structural and nutritional elements both in the farrowing and in the rearing unit, the provision of such enrichment materials should be started at a very young age. Such continuous enrichment can yield positive effects on post-weaning performance (Vanheukelom et al., 2011), but it should be noted that a pre-weaning provision has to be followed by a post-weaning provision, because any later loss of enrichment is assumed to lead to compromised welfare in the weaned piglets that were used to prior enrichment (Oostindjer et al., 2014). A decrease in mutual oral manipulations and skin lesions was detected in suckling piglets when enrichment with straw towers, paper dispensers, chew-bones and ropes was offered starting in the farrowing unit, in which the piglets began to explore the material starting on their second day of life and continuing (Fels et al., 2016). However, manipulable materials at this early stage do not seem to reduce aggressive interactions after weaning and when mixing piglets from different litters, but positive long-term effects can be assumed (Rauterberg et al., 2018). Interestingly, the use of enrichment material at this early age seems also to be linked to feed intake, because the interaction with substrates increases the flexibility towards new objects, including food, and allows the piglets to develop a broader explorative and feeding-related behaviour (Oostindjer et al., 2014). Moreover, it was shown that enrichment increases the behavioural flexibility of pigs in general and decreases fear of novel objects (Tönepöhl et al., 2012). After weaning, enrichment material can distract the piglets from performing stress-related abnormal behaviour (Fraser et al., 1991). Therefore, enrichment materials are also one set of the main preventive and emergency tools in cases of tail biting. Summarising the facts about enrichment, even the provision of small quantities of material was shown to produce positive effects (Fig. 3). Enrichment is an effective tool for increasing piglet welfare, but the substrates have to be in a proper hygienic quality because their potential to carry and transfer pathogens and toxins has already been shown; peat and material originating from maize can especially present a hygienic risk because they possibly contain mycobacteria and mycotoxins (Wagner et al., 2018).

2.3 Case study: implementation of elevated platforms in the nursery unit

One practical and inexpensive method for increasing space allowance in existing nursery pens is implementation of elevated platforms. In this way, split-level housing systems are established, which not only offer more space, but also more structure in the piglets' environment. This idea was first mentioned in the scientific literature about thirty years ago (Fraser et al., 1986; Phillips and Fraser, 1987). However, it has never been practiced on a larger scale, but has recently

Figure 3 Provision of linen straw as enrichment material in a trough, one possibility to provide substrates in nurseries with fully slatted floors (Photo: ITTN, TiHo).

drawn attention because of the ongoing debate concerning animal welfare in some European countries and the potential benefit that it possibly imparts to the piglets. Therefore, we examined the use of elevated platforms for weaners on farm and evaluated possible advantages and limitations in more detail (Fels et al., 2018b,c). In the experimental nursery unit, a platform of 5 m^2 was not only installed to provide more space per animal, but also to offer enrichment by nine different manipulable materials and toys installed on the upper level of the platform aimed to create an activity area (Fig. 4). Moreover, it was hypothesised that the additional structural elements provided by the elevated platform lead to the establishment of different functional areas, that is, for resting, activity, feeding and dunging. Forty piglets were weaned and grouped together in the two-level-pen for 33 days.

Scientific analyses included behavioural observations via video-recordings using the scan-sampling method to determine the location of the piglets within the two-level pen and their body postures (standing/sitting and lying) (Fels et al., 2018c). Moreover, ten focal animals were chosen per pen and followed continuously during the housing period to assess if they actually use the platform, and to reveal individual differences in space utilisation (Fels et al., 2018b). In order to test possible impacts on barn climate and the general hygiene, floor soiling was documented, and ammonia and air velocity were measured, too (Fels et al., 2018c). Our results showed that the piglets used the offered space to establish an activity area on the platform, a lying area under the platform, and dunging and feeding areas. This structuring led to the desired dung concentration in the dunging area; however, the animals in this study also defecated in all other areas, but to a lower extent, however, no effects of the elevated platform on the barn climate were detected (Fels

Figure 4 Enriched two-level pen in the nursery (Photo: ITTN, TiHo).

et al., 2018c). The elevated platform was used continuously during the housing period (Fig. 5), and especially in the piglets' active hours in the afternoon (Fels et al., 2018c). Each of the focal animals used the platform during the housing period; however, individual differences became apparent (Fels et al., 2018b).

The studies clearly showed that a two-level pen was a good option to increase the available space, as well as availability of enrichment and structure in existing nurseries and with those additions, piglet welfare. In our studies, the platform was offered as additional space and was not included in stocking density calculations (m^2 per animal). The legally required space was covered already by the space at the base level, and further studies might evaluate whether some recognition of the platform to fulfil (legal) spatial requirements can be accepted based on animal welfare aspects to increase the farmers' willingness to equip their nurseries with two-level pens. Full recognition of the additional space should not be recommended based on the present studies, because behavioural observations revealed that the piglets rested at night mainly under the platform and therefore, the additional space was not used in periods of inactivity (Fels et al., 2018b).

2.4 Nutritional stress: optimising feed and water provision

Sufficient water and nutrient intake is essential for welfare; otherwise, the piglet becomes impaired by thirst, hunger, and cold. At the usual age of

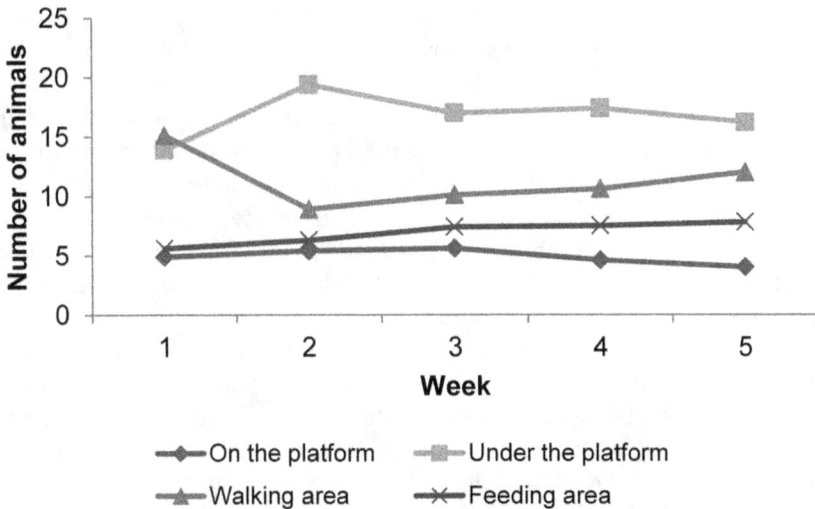

Figure 5 Mean number of animals per five minutes in the different pen areas during week 1-5, total number of 40 animals in the two-level pen (reprinted from Fels et al., Journal of Applied Animal Welfare Science 21: 267-282, 2018; with the permission of Taylor and Francis Group, www.tandfonline.com).

weaning at three to four weeks of age, sow's milk represents the main source of nutrient intake. In the farrowing unit, the sow nurses her whole litter at frequent intervals, providing them with both liquid and solid nutrients. After weaning, the piglets have to face the facts that there is no sow and no milk anymore, and they have to satisfy hunger and thirst via different, hitherto mostly unknown, resources. It takes some time to get the piglets accustomed to these new supplies, and voluntary nutrient intake decreases during that time. Bruininx et al. (2001) showed in an experiment without creep feed provision during the suckling period, that after weaning, single piglets fasted up to 50 h before they start eating, while 50% of the piglets started eating within 4 h after weaning. In another experiment, piglets without access to creep feed took longer to start feeding, up to 76 h after weaning, than piglets that got creep feed during lactation with 20 h longest before initial feed intake after weaning (Bruininx et al., 2002). Decreasing the number of animals per feeder hole from 7.5 freshly weaned piglets per feeder hole to 3.75 resulted in a decrease of average time to the onset of feed intake from 39.16 h to 31.87 h and increased feed consumption (Laskoski et al., 2019). During the critical period after weaning, in addition to changes in the gastrointestinal tract, such as crypt hyperplasia and villous atrophy, a weight loss can be observed in nearly each pig (Le Dividich and Sève, 2000; Pluske et al., 1997). Usually after two to three days longest, piglets should start feeding, but even then, they are still not gaining weight. This is caused by the fact that besides the adaptation to the technical facilities, another drawback is the slow habituation of the porcine digestive tract to the new feed composition that contains ingredients, which have to be digested by enzymes (Pluske et al., 1997). Before weaning, piglets were used to sow's milk being rich in animal protein and liquids, and their enzyme expression has to be increased in order to digest the new diet. To facilitate this process, creep feed should already be provided during the suckling period. At the time of weaning, the digestive system of the piglets is still immature, characterised by the mentioned reduced activity of enzymes, changes in intestinal morphology and reduced digestion in the small intestine (Jayaraman and Nyachoti, 2017). The metabolised energy intake (ME) in the first week post-weaning is estimated to be around 60-70% of the pre-weaning milk intake, increasing to 100% only after approximately 2-3 weeks post-weaning (Campbell et al., 2013; Pluske et al., 1997). In these first weeks in the nursery unit, the piglets' digestive system is not only limited in absorptive and digestive capacities, but also in its intestinal barrier function (Moeser and Blikslager, 2007). Therefore, cellular permeability is increased, and pathogens and toxins can cross the epithelium and cause further problems such as infections, inflammation and growth reduction (Campbell et al., 2013; Moeser and Blikslager, 2007; Oostindjer et al., 2014). This process is exacerbated by immune system activation at weaning, for instance, an upregulation of pro-inflammatory cytokines during weaning (Campbell et al., 2013). Thus, piglets are very susceptible to diarrhoea during this period, and it is no coincidence that diarrhoea is often observed during the second and third weeks post-weaning (Madec et al., 1998). Diarrhoea is not only a health, but also a welfare problem, and should therefore be prevented as much as possible.

One of the most important prevention measures is to smoothen the abrupt nutritional changes after weaning. Piglets often show food neophobia, increased by the general stressful situation at weaning (Oostindjer et al., 2014). In order to familiarise the piglets to solid feed and to trigger enzyme expression, creep feed should be offered already in the farrowing pen, as already mentioned. Piglets under commercial conditions eat small amounts of creep feed from an age of 12 days onwards (Pajor et al., 1991). At this young age, the provision of creep feed in the farrowing unit does not necessarily lead to a high intake of solid feed by the young piglets, varying individually between <1% and 20% of the daily metabolic intake (Pluske et al., 1995). However, provision of creep feed helps the piglets become accustomed to solid feed, and a positive correlation between pre- and post-weaning feed intake was shown under commercial conditions (Bruininx et al., 2002; Sulabo et al., 2010). A promising way to increase feed intake in suckling piglets, and, subsequently, in weaners is to allow the piglets to interact more with the sow and to learn to eat solid food by olfactory, visual and auditory cues learned from her (Oostindjer et al., 2014). This social learning is best when the sow and her piglets eat the same, or similar in taste feed from the same location (Oostindjer et al., 2014). Piglets interacting more with the sow in loose-housing systems pre-weaning showed more playing behaviour and less damaging behaviour in the first two weeks after weaning (Oostindjer et al., 2011)

After weaning, the same creep feed should be given in the nursery unit in order to offer familiar food to the piglets and to avoid abrupt changes. If possible, synchronal feeding should be enabled, where all piglets can access feed at the same time. Besides the provision via the feeders, small amounts of pellets should be offered on floor mats or on solid floor on the first day after weaning (Mavromichalis, 2001). In order to offer the feed via different options, mats and troughs, for instance, help to increase feed intake, because the pigs' curiosity is addressed, and they can discover different possibilities. Small quantities of pellets should be given up to three times per day to stimulate feed intake. Another approach for increasing feed intake is to offer food out of specially designed feeders that aim at the explorative behaviour in the piglets. In this way, it was shown that by offering creep feed via a 'playfeeder', which stimulated exploration, feed intake increased (Kuller et al., 2010). Placing mats in proximity to the feeder helps the piglets to get accustomed to the feeder and consume more feed in this way (Mavromichalis, 2001). These mats should be equipped with a rim to avoid rolling, pushing and wasting of the pellets. After two to three days post-weaning, the piglets should become accustomed to the feeders. It has to be kept in mind that heavy piglets at weaning are often slower in adapting to dry diets than their lighter litter mates, because they were used to drink large quantities of milk instead of eating creep feed offered in the farrowing pen. Same attention with regard to diet has to be given to all piglets. Another point to consider is that provision of a high amount of milk is important for a high feed intake capacity even later in life; therefore, supplementing the sow's diet during late gestation and lactation helps to improve piglets' performance and provides them a good start in life (Kirkden et al., 2013). Other approaches for reducing nutritional stress by the use of special feed ingredients such as prebiotics and probiotics, n-3 long chain polyunsaturated fatty acids, n-6 linoleic acid or glutamine still need to be investigated (Campbell et al., 2013). Moreover, the correlation between feed intake and other important environmental factors is of importance. Feed intake and ambient temperature are negatively correlated; however, with the weaners' higher requirements for temperature, a negative impact on feed intake, which is directly related to high ambient temperature, is rarely observed, except in very hot summer months (Dong and Pluske, 2007). Other than temperature, relative humidity does not seem to influence the performance of weaners (Noblet et al., 2001).

Concerning water provision, the piglets have to learn that after weaning water is provided from a source other than the solid feed, unlike the situation at the sow, and moreover, the way in which to use drinkers. Sufficient intake of clean, fresh and safe water is of utmost importance, affecting feed intake and therewith the growth rate. Eating and drinking times in newly weaned pigs were shown to be positively associated (Dybkjaer et al., 2006). However, it can take more than one week after weaning to reach the same intake level of fluids in the pigs as observed before weaning (Brooks and Tsourgiannis, 2003). Nipples, especially, can cause problems because their mode of use is not in line with the natural drinking position of pigs, which is done with their heads in a low position. Moreover, the delivery rate from nipple drinkers is often too low. Barber et al. (1989) described best effects on both feed intake and growth rate with water delivery rates of 0.700 L/min. Bowls and open water sources with a large reservoir capacity offer a more convenient and obvious way to drink and meet the needs of the weaners (Barber et al., 1989). As for feed, several options of water sources help the piglet to increase their intake of liquids. Concerning the numbers of drinkers per weaner, Dong and Pluske (2007) recommended a minimum of one drinker per 8–10 pigs. Drinkers for piglets should also be offered in the farrowing pen to promote water intake after weaning.

2.5 Social stress: approaches to lower welfare impacts

Social stress in weaned piglets is caused by separation from their mother and by regrouping with unfamiliar pen mates. Directly after separation, the stress caused by the separation from the mother and litter mates is mainly expressed vocally, which is in most cases only temporary. Longer-lasting is the occurrence of agonistic interactions in the new social group; directly after weaning, piglets fight to establish a new social hierarchy. Aggressive fighting is one of the most important stressors at weaning (Kutzer et al., 2009), resulting in social and physiological stress which potentially influences growth rate, causes injuries and even leads to death (Arey and Edwards, 1998). Mixing piglets, especially in combination with other adverse factors, such as high stocking density, causes stress, and if this social stress becomes chronic, it not only results in decreased growth, but also in immunosuppression and impairment of the epithelial barrier as shown in 10-week-old grower piglets (Li et al., 2017). Usually, the new rank order in a mixed group of piglets is established within 48-72 h (Fels et al., 2014a). The main reason for this increase in restriction of aggressive behaviours can be explained by the avoidance strategies of the subdominant animals (Jensen, 1982). The dominant animals are the ones that engage in more frequent fighting and more often initiate fights than the aggressors (Fels and Kemper, 2017; Rhim et al., 2015). Moreover, they often show higher weight gains by having easier access to resources (Schoenfelder, 2005). In this way, the rank order shows a direct effect on animal health and welfare. Once a social hierarchy is formed, it usually remains stable if no further animals join the group and if conditions are favourable. However, aggressive behaviours, such as dominating and threatening behaviour, can be observed even with an established rank order, when feed is restricted, animal-feeder ratio is limited, stocking density is adverse, or retreat possibilities are lacking (Fels et al., 2014b). With regard to stocking density, most fighting occurs in intermediate stocking densities, when animals cannot avoid one another, and the dominant ones can benefit from gaining access to resources by expelling other animals (Weary et al., 2008).

In order to reduce the impact of aggressive behaviours caused by post-weaning mixing and the encroaching welfare impairments, several strategies were tested, mostly without any success. These strategies only delayed the occurrence of agonistic behaviours but did not lead to suppression (Fels et al., 2014a). Distractive feeding and straw provision did not cause a steady reduction in aggression in pigs (Arey and Edwards, 1998; Arey and Franklin, 1995). Other approaches, such as the application of sedatives or deodorant substances or the reduction of light, only led to a delay in occurrence of aggression (Fels et al., 2014a). Approaches that are more promising aim at shifting mixing to an earlier stage. Under semi-natural conditions, piglets leave the nest and mix with piglets from other litters at an age of ten days (Jensen and Redbo, 1987) with only low levels of aggression. Subsequently, social relations between the piglets are established by playful, aggressive interactions, resulting in positive social integration (Petersen et al., 1989). Under farm conditions, it was shown that aggression decreases when piglets from different litters are allowed to co-mingle before weaning because the social hierarchy is formed more easily with less fighting and consequently, with fewer injuries (Morgan et al., 2014; Schrey et al., 2019; and as reviewed by Weary et al., 2008). At the age between five and twelve days, the establishment of the social hierarchy was found to be 'especially good', as reported by D'Eath (2005). These observations led to several technical approaches to enable piglets to mix with non-litter mates in the suckling period before weaning, all showing positive effects resulting in decreased aggressive behaviours and injuries, not only directly after weaning, but also longer-lasting over the fattening period (Olsson et al., 1999; Parratt et al., 2006). Such early socialised piglets seem to be better prepared for social and non-social challenges at weaning (Hillmann et al., 2003), which emphasises the importance of the early life experience for long-lasting animal welfare. Technically, the easiest and cheapest way to enact socialisation is a piglet door between neighbouring pens, thus allowing the piglets to mix and move around in a larger area than just the direct farrowing pen. Kutzer et al. (2009) showed a reduction of stress after weaning and of skin lesions due to agonistic interactions as well as a better performance in piglets with early social contact starting 10 days post-partum, via piglet doors between two adjacent pens. Another possibility is to offer a communal piglet area in which piglets of different litters can mix. Finally, group housing of lactating sows in multi-suckling systems also allows contact between the litters. Several studies have shown reduced aggression at weaning in the piglets born and raised in these systems compared to piglets from conventional single-housing pens (Bohnenkamp et al., 2013; van Nieuwamerongen et al., 2014; Schrey et al., 2019).

If a pre-weaning socialisation cannot be realised, at least mixing piglets from different litters should be avoided as far as possible. With larger group sizes in the nursery unit, this mixing becomes especially more difficult. It was shown that after weaning and mixing, aggressive behaviour is mainly observed between non-litter mates and increases with higher numbers of litters included in the group (Hwang et al., 2016; Fels and Kemper, 2017). Mixing should be minimised to decrease agonistic behaviours, but also to reduce the risk of pathogen transmission as mixing has been described as very significant risk factor for *Streptococcus suis* (Dee et al., 1993) and *Actinobacillus pleuropneumoniae* infections (Tobias et al., 2014). This point also has to be taken into consideration when pre-weaning mingling of piglets is performed.

Other behaviours directed to the pen mates include massaging, belly-nosing, mounting, or tail-biting and indicate a high level of stress (Dudink et al.,

2006). With these behaviours, compromises in welfare follow not only for the animal performing the behaviour because of high stress levels, but also for the target animal (Oostindjer et al., 2014). Concerning the time points and period of occurrence, these can vary depending on the weaning age. The younger the piglets are at weaning, the more marked the behavioural disturbances are (Prunier et al., 2010). For instance, in Germany, in which piglets are usually weaned at three or four weeks of age, peaks of tail-biting behaviour are usually seen in the second and third week after weaning (Veit et al., 2017), whereas in Sweden, tail-biting is assessed later and at lower incidences mainly in the fattening unit (Wallgren et al., 2016). In this context, the technology of segregated early weaning has to be mentioned, which is common in large pig farm enterprises with multi-site production with all-in all-out pig flow across North America. It is characterised by weaning piglets between 7 and 21 days of age (mostly between days 12–16), which are transported to isolated housings in nurseries units. Segregated early weaning aims at breaking the infection chain by utilising acquired maternal (passive) immunity from the dam before piglets develop their own active immunity in response to pathogens (von Borell, 2000). However, it was shown that increasing weaning age from 12 to 21.5 days in a multi-site swine production system significantly improved wean-to-finish growth, with weight at weaning having more influence on the post-weaning growth than changing the feeding regime (Main et al., 2004). Worobec et al. (1999) investigated the effects of very early weaning at 7 days on piglets' behaviour and performance, and found behaviour problems such as nosing and chewing pen-mates related to welfare, and lower weights at 6 weeks of age compared to piglets weaned at 14 or 28 days. They concluded that piglets do not adapt well to very early weaning at 7 days, with slightly better adaption to weaning at 14 days and significantly better adaption to weaning at 28 days (Worobec et al., 1999). Another point to consider is that weaning very early in lactation is also more difficult for the sow (Robert et al., 1999). Due to the high impact of segregated early weaning on the welfare of piglets, also including management of low-weight piglets and transportation, weaning piglets closer to 21 days of age was recommended to find both production and welfare advantages (Robert et al., 1999).

It was shown that piglets from loose-housed sows expressed less belly nosing and other manipulative post-weaning behaviours (Oostindjer et al., 2014). Of course, in this context, the environment plays a key role; piglet-directed behaviours, such as mutual oral manipulation, occur especially in barren environments and are very low in piglets raised and weaned outdoors or in other complex environments (Weary et al., 2008). Therefore, an effective way to react on or to prevent behavioural disturbances in the nursery is the provision of adequate enrichment or bedding material. Bedding, especially straw, has been shown to be a recreational stimulus with both representing a stimulus for rooting and a source of thermal and physical comfort (Fraser et al., 1991).

3 Optimised management: the human key factor

Management during the nursery stage includes batch management strategies, nutrition practices, maintenance of hygiene and biosecurity standards, and disease prevention protocols in addition to animal welfare considerations. Even if the latter are often not explicitly mentioned, careful optimisation of the other factors form a solid foundation for achieving high welfare standards if a few further above-mentioned aspects are included. However, even if all these aspects are considered, the success of a nursery, which enables animal welfare, depends on one key factor: the stockperson. The stockman is of utmost importance, because without an excellent staff, it is impossible to realise piglet welfare at the weaning and nursery stage even if all other factors are at a good level. The first weeks in the nursery unit are one of the most critical periods in a pig's life and demand most attention by the farm staff because most of the welfare problems in the nursery arise during that period. Because the sensitive piglets experience stress from different factors, as explained above, any other stressful procedures, for instance, vaccinations, should be avoided in the period around weaning to minimise additional stressors with negative impacts on health and welfare. Besides the preparations and implementations of measures previously mentioned, the stockman has to handle the piglets, to keep a careful eye on them, and decide whether and when actions have to be taken to maintain pig welfare. During and after weaning, piglets undergo handling procedures that they have previously not experienced (Orihuela et al., 2018). Thus, concerning handling, especially in weaners, all animals should be treated calmly, carefully and gently. Any kind of lesion has to be avoided by picking them up only when needed using correct picking form, which is done by the back legs and if only caught by one leg, support is needed under the chest. Moreover, they should be let down also very carefully. Having a careful eye on the piglets and their facilities is essential; feed and water intake should be checked daily in the nursery unit in addition to symptoms of diseases, such as diarrhoea and any signs of irregularities. When taking care of the piglets, attention must be given to the details, which presents a challenging and demanding task, especially for beginners. Therefore, appropriate staff training on a regular basis and an exchange of experiences are essential. Even though new trends, especially in the field of digitalisation, offer innovative ways to monitor the animals, including camera systems that detect behavioural patterns, such as increased movements, at this delicate age in the nursery, human care cannot be replaced.

4 Conclusion

Optimising piglet welfare at the nursery stage is one of the most difficult tasks in modern pig farming. In the existing systems and management structures, the young age of the animals at weaning in most countries represents a big challenge, and the interaction of different approaches has to work smoothly to fulfil the most urgent needs of the piglets. In order to implement weaner welfare, the farmers have to be aware that every step towards best welfare, including health, is always also one step towards good performance. Any drawbacks, but also any benefits, are reflected in the balance sheet. Weaners are future capital and have to be treated in a respectful manner by doing everything possible to avoid or at least reduce environmental, nutritional and social stress. In the myriads of scientific studies, manifold aspects of improving weaners' performance and welfare in existing systems have been elucidated. The challenge lies more in the actual combination and transfer of knowledge into practice, including scientific support to analyse the outcome. This can only be realised by on-farm research projects, a highly motivated and skilled farm staff, and incentives for those farms that successfully improve welfare. Whether changes in the whole pig farming framework towards more animal friendly systems, for instance, with longer lactation periods and higher age at weaning, as already realised in some countries, will happen on a broader base in the near future is doubtful. It is therefore all the more important to make every effort within the realms of possibility and considering all the important points mentioned in this chapter to enable good pig welfare during the nursery stage.

5 Future trends in research

Even though a large number of studies have focussed on weaners' welfare and possible ways to improve it in commercial pig production over the last decades, there is still room for improvements. Future trends in research might depend on the demands and settings in the respective countries. In Northern and Middle European countries, where animal welfare is a highly relevant topic, and changes in the husbandry systems can be expected in future, open research questions are, for instance, the evaluation of 'enriched' nurseries with more structures, bedded floors and outdoor access, and the possible consequences on welfare and production parameters. In other countries, where the existing systems are not expected to change, other topics such as the optimised transportation of freshly, maybe early, weaned piglets may have priority. As a general topic, the use of advanced computer-based, digitalized technologies is most probably a huge future research trend in the nursery, as well. Automated techniques, for example, for detecting aggressive episodes in weaned pigs (e.g. Chen et al., 2020), may be beneficial for increasing weaners' welfare. Concluding, research on optimising welfare in nurseries has to consider many factors from different areas, including behaviour studies, process management and animal hygiene, and therefore, is a major challenge.

6 Where to look for further information

- Homepage of the EU-Project FareWellDock (tail docking, tail biting, health and enrichment in pigs): https://farewelldock.eu/info/.
- Homepage of the EU-Project GroupHouseNet (synergy for preventing damaging behaviour in group-housed pigs and chickens): https://www.grouphousenet.eu/.
- Homepage of the EU PiG Innovation group: https://www.eupig.eu/.
- Homepage of the European Reference Centre for Animal Welfare Pigs (EURCAW-Pigs): https://www.eurcaw.eu/en/eurcaw-pigs.htm.

7 References

Aarnink, A. J. A., Swierstra, D., van den Berg, A. J. and Speelman, L. (1997). Effect of type of slatted floor and degree of fouling of solid floor on ammonia emission rates from fattening piggeries. *Journal of Agricultural Engineering Research* 66(2): 93-102.

Arey, D. S. and Edwards, S. A. (1998). Factors influencing aggression between sows after mixing and the consequences for welfare and production. *Livestock Production Science* 56(1): 61-70.

Arey, D. S. and Franklin, M. F. (1995). Effects of straw and unfamiliarity on fighting between newly mixed growing pigs. *Applied Animal Behaviour Science* 45(1-2): 23-30.

Averós, X., Herranz, A., Sánchez, R. and Gosálvez, L. F. (2009). Effect of the duration of commercial journeys between rearing farms and growing-finishing farms on the physiological stress response of weaned piglets. *Livestock Science* 122(2-3): 339-344.

Averós, X., Knowles, T. G., Brown, S. N., Warriss, P. D. and Gosálvez, L. F. (2010). Factors affecting the mortality of weaned piglets during commercial transport between farms. *Veterinary Record* 167(21): 815-819.

Barber, J., Brooks, P. H. and Carpenter, J. L. (1989). The effects of water delivery rate on the voluntary food intake, water use and performance of early-weaned pigs from 3 to 6 weeks of age. *BSAP Occasional Publication* 13: 103-104.

Beattie, V., Walker, N. and Sneddon, I. (1995). Effects of environmental enrichment on behaviour and productivity of growing pigs. *Animal Welfare* 4: 207-220.

Beattie, V. E., Walker, N. and Sneddon, I. A. (1996). An investigation of the effect of environmental enrichment and space allowance on the behaviour and production of growing pigs. *Journal of Applied Animal Behaviour Science* 48(3-4): 151-158.

Berry, R. J. and Lewis, N. J. (2001). The effect of duration and temperature of simulated transport on the performance of early-weaned piglets. *Canadian Journal of Animal Science* 81(2): 199-204.

Bøe, K. (1991). The process of weaning in pigs: when the sow decides. *Applied Animal Behaviour Science* 30(1-2): 47-59.

Bohnenkamp, A. L., Traulsen, I., Meyer, C., Müller, K. and Krieter, J. (2013). Comparison of growth performance and agonistic interaction in weaned piglets of different weight classes from farrowing systems with group or single housing. *Animal* 7(2): 309-315.

Bracke, M. B. M., Metz, J. H. M., Spruijt, B. M. and Schouten, W. G. P. (2002a). Decision support system for overall welfare assessment in pregnant sows B: validation by expert opinion. *Journal of Animal Science* 80(7): 1835-1845.

Bracke, M. B. M., Spruijt, B. M., Metz, J. H. M. and Schouten, W. G. P. (2002b). Decision support system for overall welfare assessment in pregnant sows A: model structure and weighting procedure1. *Journal of Animal Science* 80(7): 1819-1834.

Brooks, P. and Tsourgiannis, C. A. (2003). Factors affecting the voluntary feed intake of the weaned pig. In: Pluske, J., Le Dividich, J. and Verstegen, M. W. A. (Eds), *Weaning the Pig: Concepts and Consequences.* Wageningen Academic Publishers, Wageningen, The Netherlands, 81-116.

Bruininx, E. M. A. M., Binnendijk, G. P., van der Peet-Schwering, C. M. C, Schrama, J. W., den Hartog, L. A., Everts, H. and Beynen, A. C. (2002). Effect of creep feed consumption on individual feed intake characteristics and performance of group-housed weanling pigs. *Journal of Animal Science* 80: 1413-1418.

Bruininx, E. M. A. M., van der Peet-Schwering, C. M. C., Schrama, J. W., Vereijken, P. F. G., Vesseur, P. C., Everts, H., den Hartog, L. A. and Beynen, A. C. (2001). Individually measured feed intake characteristics and growth performance of group-housed weanling pigs: effects of sex, initial body weight, and body weight distribution within groups. *Journal of Animal Science* 79: 301-308.

Brumm, M. C., Ellis, M., Johnston, L. J., Rozeboom, D. W., Zimmerman, D. R. and NCR-89 Committee on Swine Management (2001). Interaction of swine nursery and grow-finish space allocations on performance. *Journal of Animal Science* 79: 1967-1972.

Brumm, M. C. and Gonyou, H. W. (2001). Effects of facility design on behavior and feed and water intake. In: Lewis, A. J. and Southern, L. L. (Eds), *Swine Nutrition* (2nd edn). CRC Press LLC Press, Boca Raton, FL, 499-517.

Campbell, J. M., Crenshaw, J. D. and Polo, J. (2013). The biological stress of early weaned piglets. *Journal of Animal Science and Biotechnology* 4(1): 19.

Chen, C., Zhu, W., Steibel, J., Siegford, J., Wurtz, K., Han, J. and Norton, T. (2020). Recognition of aggressive episodes of pigs based on convolutional neural network and long short-term memory. *Computers and Electronics in Agriculture* 169: 105166.

Cooper, J. E. (1975). The use of the pig as an animal model to study problems associated with low birthweight. *Laboratory Animals* 9(4): 329-336.

Council Directive 2008/120/EC (2008). *Of 18 December 2008 Laying down Minimum Standards for the Protection of Pigs.* https://eur-lex.europa.eu/legal-content/EN/AL L/?uri=CELEX%3A32008L0120. Accessed November 14, 2019.

D'Eath, R. B. (2005). Socialising piglets before weaning improves social hierarchy formation when pigs are mixed post-weaning. *Applied Animal Behaviour Science* 93(3-4): 199-211.

Dee, S. A., Carlson, A. R., Winkelman, N. L. and Corey, M. M. (1993). Effect of management practices on the Streptococcus suis carrier rate in nursery swine. *Journal of the American Veterinary Medical Association* 203(2): 295-299.

De Vos, M., Che, L., Huygelen, V., Willemen, S., Michiels, J., Van Cruchten, S. and Van Ginneken, C. (2014). Nutritional interventions to prevent and rear low birth weight piglets. *Journal of Animal Physiology and Animal Nutrition* 98(4): 609-619.

Done, S. H., Chennells, D. J., Gresham, A. C. J., Williamson, S., Hunt, B., Taylor, L. L., Bland, V., Jones, P., Armstrong, D., White, R. P., Demmers, T. G. M., Teer, N. and Wathes, C. M. (2005). Clinical and pathological responses of weaned pigs to atmospheric ammonia and dust. *Veterinary Record* 157(3): 71-80.

Dong, G. Z. and Pluske, J. R. (2007). The low feed intake in newly-weaned pigs: problems and possible solutions. *Asian-Australasian Journal of Animal Sciences* 20(3): 440-452.

Dudink, S., Simonse, H., Marks, I., de Jonge, F. H. and Spruijt, B. M. (2006). Announcing the arrival of enrichment increases play behaviour and reduces weaning-stress-induced behaviours of piglets directly after weaning. *Applied Animal Behaviour Science* 101(1-2): 86-101.

Dybkjaer, L., Jacobsen, A. P., Togersen, F. A. and Poulsen, H. D. (2006). Eating and drinking activity of newly weaned piglets: effects of individual characteristics, social mixing, and addition of extra zinc to the feed. *Journal of Animal Science* 84(3): 702-711.

EFSA Panel on Animal Health and Welfare (2005). *EFSA Journal* 3(10): 268, doi:10.2903/j.efsa.2005.268.

Fels, M., Giersberg, M., Bill, J., Gillandt, K. and Kemper, N. (2016). Effects of a uniform environmental enrichment during lactation and after weaning on the behaviour and skin lesion score of piglets. *Züchtungskunde* 88: 241-253.

Fels, M., Hartung, J. and Hoy, S. (2014a). Social hierarchy formation in piglets mixed in different group compositions after weaning. *Applied Animal Behaviour Science* 152: 17-22.

Fels, M., Thays Sonoda, L., Rauterberg, S., Hartung, J. and Kemper, N. (2014b). Aggressive behaviour of piglets after mixing - causes, factors and control measures. *Der Praktische Tierarzt* 95: 952-960.

Fels, M. and Kemper, N. (2017). Fighting activity and establishment of a litter-associated dominance order in weaned piglets after mixing. *Berliner und Münchener Tierärztliche Wochenschrift* 130: 306-313.

Fels, M., Konen, K., Hessel, E. and Kemper, N. (2018a). Determination of static space occupied by individual weaner and growing pigs using an image-based monitoring system. *The Journal of Agricultural Science* 156(2): 282-290.

Fels, M., Lüthje, F., Bill, J., Aleali, K. and Kemper, N. (2018b). Elevated platforms for weaners: do pigs use the extra space? *Veterinary Record* 183(7): 222-222.

Fels, M., Lüthje, F., Faux-Nightingale, A. and Kemper, N. (2018c). Use of space and behavior of weaned piglets kept in enriched two-level housing system. *Journal of Applied Animal Welfare Science* 21(3): 267-282.

Fels, M., Konen, K., Hessel, E. and Kemper, N. (2019). Biometric measurement of static space required by weaned piglets kept in groups of eight during 6 weeks. *Animal Production Science* 59(7): 1327-1335.

Fraser, D. (1985). Selection of bedded and unbedded areas by pigs in relation to environmental temperature and behaviour. *Applied Animal Behaviour Science* 14(2): 117-126.

Fraser, D., Phillips, P. A. and Thompson, B. K. (1986). A test of a free-access two-level pen for fattening pigs. *Journal of Animal Science* 42(2): 269-274.

Fraser, D., Phillips, P. A., Thompson, B. K. and Tennessen, T. (1991). Effect of straw on the behaviour of growing pigs. *Applied Animal Behaviour Science* 30(3-4): 307-318.

Gondret, F., Lefaucheur, L., Louveau, I., Lebret, B., Pichodo, X. and Le Cozler, Y. (2005). Influence of piglet birth weight on postnatal growth performance, tissue lipogenic capacity and muscle histological traits at market weight. *Livestock Production Science* 93(2): 137-146.

Hales, J., Moustsen, V. A., Nielsen, M. B. F. and Hansen, C. F. (2013). Individual physical characteristics of neonatal piglets affect preweaning survival of piglets born in a noncrated system. *Journal of Animal Science* 91(10): 4991-5003.

Hennig-Pauka, I., Menzel, A., Boehme, T. R., Schierbaum, H., Ganter, M. and Schulz, J. (2019). Haptoglobin and C-reactive protein–non-specific markers for nursery conditions in swine. *Frontiers in Veterinary Science* 6: 92.

Herpin, P., Bertin, R., Le Dividich, J. and Portet, R. (1987). Some regulatory aspects of thermogenesis in cold-exposed piglets. *Comparative Biochemistry and Physiology. A, Comparative Physiology* 87(4): 1073-1081.

Hillmann, E., von Hollen, F., Bünger, B., Todt, D. and Schrader, L. (2003). Farrowing conditions affect the reactions of piglets towards novel environment and social confrontation at weaning. *Applied Animal Behaviour Science* 81(2): 99-109.

Hwang, H. S., Lee, J. K., Eom, T. K., Son, S. H., Hong, J. K., Kim, K. H. and Rhim, S. J. (2016). Behavioral characteristics of weaned piglets mixed in different groups. *Asian-Australasian Journal of Animal Sciences* 29(7): 1060-1064.

Jayaraman, B. and Nyachoti, C. M. (2017). Husbandry practices and gut health outcomes in weaned piglets: a review. *Animal Nutrition* 3(3): 205-211.

Jensen, P. (1982). An analysis of agonistic interaction patterns in group-housed dry sows–aggression regulation through an "avoidance order". *Journal of Applied Animal Ethology* 9(1): 47-61.

Jensen, P. (1986). Observations on the maternal behaviour of free-ranging domestic pigs. *Applied Animal Behaviour Science* 16(2): 131-142.

Jensen, P. and Redbo, I. (1987). Behavior during nest-leaving in free-ranging domestic pigs. *Applied Animal Behaviour Science* 110: 355-362.

Jones, J. B., Burgess, L. R., Webster, A. J. F. and Wathes, C. M. (1996). Behavioural responses of pigs to atmospheric ammonia in a chronic choice test. *Animal Science* 63(3): 437-445.

KilBride, A., Gillman, C., Ossent, P. and Green, L. (2009). Impact of flooring on the health and welfare of pigs. *In Practice* 31(8): 390-395.

Kirkden, R. D., Broom, D. M. and Andersen, I. L. (2013). Invited review: piglet mortality: management solutions. *Journal of Animal Science* 91(7): 3361-3389.

Koketsu, Y., Tani, S. and Iida, R. (2017). Factors for improving reproductive performance of sows and herd productivity in commercial breeding herds. *Porcine Health Management* 3: 1.

Kuller, W. I., Tobias, T. J. and Van Nes, A. (2010). Creep feed intake in unweaned piglets is increased by exploration stimulating feeder. *Journal of Livestock Science* 129(1-3): 228-231.

Kutzer, T., Bünger, B., Kjaer, J. B. and Schrader, L. (2009). Effects of early contact between non-littermate piglets and of the complexity of farrowing conditions on social behavior and weight gain. *Applied Animal Behaviour Science* 121(1): 16-24.

Lallès, J. P., Boudry, G., Favier, C., Le Floch, N., Luron, I., Montagne, L., Oswald, I. P., Pié, S., Piel, C. and Sève, B. (2004). Gut function and dysfunction in young pigs: physiology. *Animal Research* 53(4): 301-316.

Laskoski, F., Faccin, J. E. G., Vier, C. M., Gonçalves, M. A. D., Orlando, U. A. D., Kummer, R., Mellagi, A. P. G., Bernardi, M. L., Wentz, I. and Bortolozzo, F. P. (2019). Effects of pigs per feeder hole and group size on feed intake onset, growth performance, and ear and tail lesions in nursery pigs with consistent space allowance. *Journal of Swine Health and Production* 27(1): 12-18.

Le Dividich, J. and Herpin, P. (1994). Effects of climatic conditions on the performance, metabolism and health status of weaned piglets: a review. *Livestock Production Science* 38(2): 79-90.

Le Dividich, J. and Sève, B. (2000). Effects of underfeeding during the weaning period on growth, metabolism, and hormonal adjustments in the piglet. *Domestic Animal Endocrinology* 19(2): 63-74.

Le Floch, N., Jondreville, C., Matte, J. J. and Seve, B. (2006) Importance of sanitary environment for growth performance and plasma nutrient homeostasis during the post-weaning period in piglets. *Archives of Animal Nutrition* 60(1): 23-34, doi:10.1080/17450390500467810.

Levis, D. G. (1997). Effect of early weaning on sow reproductive performance - a review. *Nebraska Swine Reports* 204. http://digitalcommons.unl.edu/coopext_swine/204.

Lewis, N. J. (2008). Transport of early weaned piglets. *Applied Animal Behaviour Science* 110(1-2): 128-135.

Lewis, N. J. and Berry, R. J. (2006). Effects of season on the behaviour of early-weaned piglets during and immediately following transport. *Applied Animal Behaviour Science* 100(3-4): 182-192.

Li, Y., Song, Z., Kerr, K. A. and Moeser, A. J. (2017). Chronic social stress in pigs impairs intestinal barrier and nutrient transporter function, and alters neuro-immune mediator and receptor expression. *PLoS ONE* 12(2): e0171617.

Madec, F., Bridoux, N., Bounaix, S. and Jestin, A. (1998). Measurement of digestive disorders in the piglet at weaning and related risk factors. *Preventive Veterinary Medicine* 35(1): 53-72.

Madec, F., Le Dividich, J., Pluske, J. R. and Verstegen, M. W. A. (2003). Environmental requirements and housing of the weaned pig. In: Pluske, J., Le Dividich, J. and Verstegen, M. W. A. (Eds), *Weaning the Pig: Concepts and Consequences.* Wageningen Academic Publishers, Wageningen, The Netherlands, 337-360.

Main, R. G., Dritz, S. S., Tokach, M. D., Goodbandt, R. D. and Nelssen, J. L. (2004). Increasing weaning age improves pig performance in a multisite production system. *Journal of Animal Science* 82(5): 1499-1507.

Martin, J. E., Ison, S. H. and Baxter, E. M. (2015). The influence of neonatal environment on piglet play behaviour and post-weaning social and cognitive development. *Applied Animal Behaviour Science* 163: 69-79.

Mavromichalis, I. (2001). Management of the nursery pig. *Illinois Porknet Papers*. http://livestocktrail.illinois.edu/uploads/porknet/papers/ManagementOfTheNurseryPig.PDF. Accessed October 12, 2019.

Moeser, A. J. and Blikslager, A. T. (2007). Mechanisms of porcine diarrheal disease. *Journal of the American Veterinary Medical Association* 231(1): 56-67.

Moeser, A. J., Pohl, C. S. and Rajput, M. (2017). Weaning stress and gastrointestinal barrier development: implications for lifelong gut health in pigs. *Animal Nutrition* 3(4): 313-321.

Morgan, T., Pluske, J., Miller, D., Collins, T., Barnes, A. L., Wemelsfelder, F. and Fleming, P. A. (2014). Socialising piglets in lactation positively affects their post-weaning behaviour. *Journal of Applied Animal Behaviour Science* 158: 23-33.

Muirhead, M. R., Alexander, T. J. L. and Carr, J. (2013). *Managing Pig Health: A Reference for the Farm* (2nd edn). 5M PUB, Sheffield, England.

Munsterhjelm, C., Peltoniemi, O. A. T., Heinonen, M., Hälli, O., Karhapää, M. and Valros, A. (2009). Experience of moderate bedding affects behaviour of growing pigs. *Applied Animal Behaviour Science* 118(1–2): 42–53.

Newberry, R. C. and Wood-Gush, D. G. M. (1985). The suckling behaviour of domestic pigs in a semi-natural environment. *Behaviour* 95(1–2): 11–25.

Niekamp, S. R., Sutherland, M. A., Dahl, G. E. and Salak-Johnson, J. L. (2007). Immune response of piglets to weaning stress: impacts of photoperiod. *Journal of Animal Science* 85(1): 93–100.

Noblet, J., Le Dividich, J. and Van Milgen, J. (2001). Thermal environment and swine nutrition. In: Lewis, A. J. and Southern, L. L. (Eds), *Swine Nutrition* (2nd edn). CRC Press LLC Press, Boca Raton, FL, 519–544.

Oh, H. K., Choi, H. B., Ju, W. S., Chung, C. S. and Kim, Y. Y. (2010). Effects of space allocation on growth performance and immune system in weaning pigs. *Livestock Science* 132(1–3): 113–118.

Olsson, I. A. S., de Jonge, F. H., Schuurman, T. and Helmond, F. A. (1999). Poor rearing conditions and social stress in pigs: repeated social challenge and the effect on behavioural and physiological responses to stressors. *Behavioural Processes* 46(3): 201–215.

Oostindjer, M., Bolhuis, J. E., Mendl, M., Held, S., Gerrits, W., Van den Brand, H. and Kemp, B. (2010). Effects of environmental enrichment and loose housing of lactating sows on piglet performance before and after weaning. *Journal of Animal Science* 88(11): 3554–3562.

Oostindjer, M., Kemp, B., van den Brand, H. and Bolhuis, J. E. (2014). Facilitating 'learning from mom how to eat like a pig' to improve welfare of piglets around weaning. *Applied Animal Behaviour Science* 160: 19–30.

Oostindjer, M., van den Brand, H., Kemp, B. and Bolhuis, J. E. (2011). Effects of environmental enrichment and loose housing of lactating sows on piglet behaviour before and after weaning. *Journal of Applied Animal Behaviour Science* 134(1–2): 31–41.

Orihuela, A., Mota-Rojas, D., Velarde, A., Strappini, A., de la Vega, L., Borderas, F. and Alonso-Spilsbury, M. (2018). Environmental enrichment to improve behaviour in farm animals. CABI. England. *CAB Reviews Perspectives in Agriculture Veterinary Science Nutrition and Natural Resources* 13: 1–25.

Pajor, E. A., Fraser, D. and Kramer, D. L. (1991). Consumption of solid food by suckling pigs: individual variation and relation to weight gain. *Applied Animal Behaviour Science* 32(2–3): 139–155.

Parratt, C. A., Chapman, K. J., Turner, C., Jones, P. H., Mendl, M. T. and Miller, B. G. (2006). The fighting behaviour of piglets mixed before and after weaning in the presence or absence of a sow. *Journal or Applied Animal Behavioural Science* 101(1–2): 54–67.

Petersen, H. V., Vestergaard, K. and Jensen, P. (1989). Integration of piglets into social groups of free-ranging domestic pigs. *Applied Animal Behaviour Science* 23(3): 223–236.

Phillips, P. and Fraser, D. (1987). Design, cost and performance of a free-access, two-level pen for growing-finishing pigs. *Canadian Agricultural Engineering* 29.

Pluske, J. R., Hampson, D. J. and Williams, I. H. (1997). Factors influencing the structure and function of the small intestine in the weaned pig: a review. *Livestock Production Science* 51(1–3): 215–236.

Pluske, J. R., Williams, I. H. and Aherne, F. X. (1995). Nutrition of the neonatal pig. In: Varlex, M. A. (Ed), *The Neonatal Pig: Development and Survival*. CAB Int., Wallingford (UK), 187–235.

Prunier, A., Heinonen, M. and Quesnel, H. (2010). High physiological demands in intensively raised pigs: impact on health and welfare. *Animal* 4(6): 886-898.

Quiniou, N., Dagorn, J. and Gaudré, D. (2002). Variation of piglets' birth weight and consequences on subsequent performance. *Livestock Production Science* 78(1): 63-70.

Rauterberg, S., Kemper, N. and Fels, M. (2018). Influence of environmental enrichment during lactation and after weaning on aggressive behaviour and skin lesion score of piglets. *Berliner und Münchener Tierärztliche Wochenschrift* 131(1–2): 12-19.

Rhim, S. J., Son, S. H., Hwang, H. S., Lee, J. K. and Hong, J. K. (2015). Effects of mixing on the aggressive behavior of commercially housed pigs. *Asian-Australasian Journal of Animal Sciences* 28(7): 1038-1043.

Robert, S., Weary, D. M. and Gonyou, H. (1999). Segregated early weaning and welfare of piglets. *Journal of Applied Animal Welfare Science* 2(1): 31-40.

Schoenfelder, A. (2005). The effect of rank order on feed intake and growth of fattening pigs. *DTW. Deutsche Tierarztliche Wochenschrift* 112(6): 215-218.

Schrey, L., Kemper, N. and Fels, M. (2019). Behaviour and skin injuries of piglets originating from a novel group farrowing system before and after weaning. *Agriculture*, MDPI 9: 14.

Schulz, J., Ruddat, I., Hartung, J., Hamscher, G., Kemper, N. and Ewers, C. (2016). Antimicrobial-resistant Escherichia coli survived in dust samples for more than 20 years. *Frontiers in Microbiology* 7: 866.

Scipioni, R., Martelli, G. and Antonella Volpelli, L. (2009). Assessment of welfare in pigs. *Italian Journal of Animal Science* 8(sup1): 117-137.

Smith, A. L., Stalder, K. J., Serenius, T. V., Baas, T. J. and Mabry, J. W. (2007). Effect of piglet birth weight on weights at weaning and 42 days post weaning. *Journal of Swine Health and Production* 15(4): 213-218.

Smith, L. F., Beaulieu, A. D., Patience, J. F., Gonyou, H. W. and Boyd, R. D. (2004). The impact of feeder adjustment and group size-floor space allowance on the performance of nursery pigs. *Journal of Swine Health and Production* 12(3): 111-118.

Spurlock, M. E. (1997). Regulation of metabolism and growth during immune challenge: an overview of cytokine function. *Journal of Animal Science* 75(7): 1773-1783.

Stafford, K. J. (2010). Tail biting: an important and undesirable behaviour of growing pigs. *The Veterinary Journal* 186(2): 131-132.

Stojanac, N., Stevančević, O., Potkonjak, A., Savić, B., Stančić, I. and Vračar, V. (2014). The impact of space allowance on productivity performance and Salmonella spp. shedding in nursery pigs. *Livestock Science* 164: 149-153.

Sulabo, R. C., Tokach, M. D., Dritz, S. S., Goodband, R. D., DeRouchey, J. M. and Nelssen, J. L. (2010). Effects of varying creep feeding duration on the proportion of pigs consuming creep feed and neonatal pig performance. *Journal of Animal Science* 88(9): 3154-3162.

Sutherland, M. A., Backus, B. L. and McGlone, J. J. (2014). Effects of transport at weaning on the behavior, physiology and performance of pigs. *Animals* 4(4): 657-669; doi:10.3390/ani4040657.

Sutherland, M. A., Bryer, P. J., Davis, B. L. and McGlone, J. J. (2009). Space requirements of weaned pigs during a 60 minute transport in summer. *Journal of Animal Science* 87(1): 363-370.

Tobias, T. J., Bouma, A., van den Broek, J., van Nes, A., Daemen, A. J. J. M., Wagenaar, J. A., Stegeman, J. A. and Klinkenberg, D. (2014). Transmission of Actinobacillus pleuropneumoniae among weaned piglets on endemically infected farms. *Preventive Veterinary Medicine* 117(1): 207-214.

Tönepöhl, B., Appel, A. K., Welp, S., Voss, B., König von Borstel, U. and Gauly, M. (2012). Effect of marginal environmental and social enrichment during rearing on pigs' reactions to novelty, conspecifics and handling. *Applied Animal Behaviour Science* 140(3-4): 137-145.

van Nieuwamerongen, S. E., Bolhuis, J. E., van der Peet-Schwering, C. M. and Soede, N. M. (2014). A review of sow and piglet behaviour and performance in group housing systems for lactating sows. *Animal* 8(3): 448-460.

Vanheukelom, V., Driessen, B., Maenhout, D. and Geers, R. (2011). Peat as environmental enrichment for piglets: the effect on behaviour, skin lesions and production results. *Applied Animal Behaviour Science* 134(1-2): 42-47.

Veit, C., Büttner, K., Traulsen, I., Gertz, M., Hasler, M., Burfeind, O., große Beilage, E. and Krieter, J. (2017). The effect of mixing piglets after weaning on the occurrence of tail-biting during rearing. *Livestock Science* 201: 70-73.

von Borell, E. (2000). Welfare assessment of segregated early weaning (SEW) - a review. *Archives of Animal Breeding* 43: 337-345.

Wagner, K. M., Schulz, J. and Kemper, N. (2018). Examination of the hygienic status of selected organic enrichment materials used in pig farming with special emphasis on pathogenic bacteria. *Porcine Health Management* 4: 24-24.

Wallgren, T., Westin, R. and Gunnarsson, S. (2016). A survey of straw use and tail biting in Swedish pig farms rearing undocked pigs. *Acta Veterinaria Scandinavica* 58(1): 84.

Wamnes, S., Lewis, N. J. and Berry, R. J. (2006). The performance of early-weaned piglets following transport: effect of season and weaning weight. *Canadian Journal of Animal Science* 86(3): 337-343.

Weary, D. M., Jasper, J. and Hötzel, M. J. (2008). Understanding weaning stress. *Applied Animal Behaviour Science* 110(1-2): 24-41.

Worobec, E. K., Duncan, I. J. H. and Widowski, T. M. (1999). The effects of weaning at 7, 14 and 28 days on piglet behaviour. *Applied Animal Behaviour Science* 62(2-3): 173-182.

Chapter 3

Welfare of weaned piglets

Arlene Garcia and John J. McGlone, Texas Tech University, USA

1 Introduction

Nearly 1 billion pigs are in inventory on farms around the world. About half the pigs are found in China alone. The other regions with significant number of pigs include South East Asia, Japan, Brazil, Mexico, the United States, Canada and most of Europe. Among these geographical locations, we find three general types of production models: (1) modern, industrialized production systems [Modern]; (2) small-scale low-input, often-backyard, family operations [Backyard]; and (3) natural, organic, antibiotic-free, GMO-free systems [Natural]. Modern industrialized systems have been called factory farms by those that wish to criticize keeping pigs in buildings and pens. However, pigs were moved from mud lots to buildings to improve their welfare and their health in particular. Today, we may find very low pre-weaning mortality and high health among pigs in many industrialized systems. Likewise, some presumably high-welfare systems have high pre-weaning piglet mortality and significant numbers of sows with scratches and wounds compared with modern US industrialized systems (McGlone, 2006). We have also reported good welfare among sows in both indoor and outdoor production systems that are well managed (Johnson et al., 2001). One cannot support the idea that industrialized production systems are inherently bad for pig welfare. Likewise, one cannot assume that because sows and piglets are in more natural settings that their welfare is automatically good.

http://dx.doi.org/10.19103/AS.2017.0013.24

Good and poor welfare are found both on industrialized and more on natural production systems.

Concerns and demands for improved animal welfare and animal handling systems from authorities, non-government organizations, markets and the public in general are increasing and are of utmost importance to the meat industry (Stoier et al., 2016). Today in the United States, all Modern farms must have an animal welfare audit conducted by a third party (one not associated with the farm or the market chain). Also, many Natural systems have an animal welfare assessment requirement of some kind. Backyard production systems often are not regulated or assessed.

In Europe, some farms use third-party audit programmes and some individual European countries have government oversight of pig farms. In developing countries, animal welfare audits are initiated, but they are far from comprehensive and they do not cover the majority of pigs in commercial production. We expect this to change over time. While governments vary in animal welfare requirements, multinational retail buyers of pork products now require third-party oversight of farm animal welfare in developed countries and they are slowly advancing animal welfare audit programmes in developing countries.

Requiring animal welfare third-party assessment is one step towards assuring a minimum animal welfare standard is met in the meat supply chain. However, the situation is made complex because we are unsure, in some cases, how to create the best animal welfare. Among the phases of pork production with opportunities to improve welfare, the weaning period has the greatest opportunity to positively impact most animals. By weaning period we mean from birth through a week or two after weaning. Pre-weaning mortality of piglets has dropped from over 25% 30 years ago to less than 10% today on Modern farms. Even at 10% pre-weaning mortality, over 100 million piglets die before weaning each year. The general classification of opportunities to improve piglet welfare falls in these categories (in order of the numbers of piglets impacted and the severity of the welfare issue):

- Pre-weaning mortality
- Weaning stress
- Painful practices
- Transportation

It is important to understand that poor management can lead to many welfare issues, and educating caregivers to make good judgements can greatly improve these issues. This chapter will focus on current practices that can be detrimental to piglet well-being, alternatives and/or solutions to these, and advances in technology that could improve animal well-being, profitability and sustainability.

2 Pre-weaning mortality

The causes of pre-weaning mortality are many. In an early, classic paper from 1971, Fahmy and Bernard documented the causes of mortality among 6,890 Yorkshire piglets from birth to 20 weeks of age. Mortality was 25.6% on average. Of this total mortality, 16.4% was from birth to weaning at 8 weeks of age. The causes of death among piglets were general weakness, crushing by the sow (these two comprise nearly half of all piglet deaths), scours, paralysis, rickets, anatomical abnormalities, pneumonia and anaemia. Interestingly, they reported again (in agreement with authors in the 1940s) that (1) sows with larger litters had more pre-weaning piglet deaths, (2) inbreeding of these purebred sows increased piglet mortality and (3) smaller body weight piglets have a higher risk of death. By examining Fahmy and Bernard's (1971) study on causes of piglet death, one can realize that some problems are solved on modern farms – we have less scours and pneumonia on high health farms, no inbreeding, no rickets and we have bred against weak piglets and anatomical problems. So today, around the world, 45 years later, authors confirm that larger litters have more smaller, at-risk piglets and that crushing is a major cause of pre-weaning mortality (e.g. Thailand: Nuntapaitoon and Tummaruk, 2015; Sweden: Westin et al., 2015; Kilbride et al., 2015).

Solutions to reduce pre-weaning mortality have focused on the farrowing environment, sow genetics and management practices such as cross-fostering piglets among sows.

2.1 The farrowing environment

The farrowing environment is a housing or penning or keeping system where sows reside from a few days before farrowing through weaning. It may be a single system or a combination of systems. The most common system in Modern pork production systems is the farrowing crate in which a single sow is kept in an area that does not allow her to turn around and requires that she lie down more gently along bars (often metal). The farrowing crate prevents sows from lying down quickly and crushing piglets. While the farrowing crate saves piglet lives compared with the older-style pen, it is not the only system to reduce pre-weaning mortality.

The farrowing crate was first patented in the United States in 1963 by Ingvald Eide (patent publication date Feb 19, 1963; US 3077861), but it was in widespread use before then in the United States and Europe. The first scientific publication that showed a reduction in pre-weaning mortality was by Robertson et al. (1966) in England. The farrowing crate quickly spread among Modern pig farms such that by 1980, it was the most common system in pork production in North America and Europe. Still, people criticized the farrowing crate for restricting the sow's movement. Systems were sought that allowed sows to turn around, but that did not cause a rise in piglet mortality.

The first published viable alternative to the farrowing crate was the turn-around pen. In this system, the pen is 21% larger (due to 21% larger length) and the same width as a standard farrowing crate (2.7 m × 1.95 m) and had a V-shape to the farrowing railings. McGlone and Blecha (1987) showed this system had even lower pre-weaning mortality than the standard farrowing crate. The turn-around crate was modelled after a pork producer in Minnesota who used what was called the Compart crate (which was actually a pen). McGlone and Morrow-Tesch (1990) showed that the sloped-floor farrowing pen also had lower pre-weaning mortality as a farrowing crate. Frank Hurnick from Canada developed an ellipsoid farrowing crate that equalled the standard farrowing crate in piglet mortality (Lou and Hurnick, 1994). Greg Cronin developed what he called the Werribee Pen that had equal pre-weaning mortality as the farrowing crate (Cronin et al., 2000). Outdoor farrowing huts vary significantly in pre-weaning mortality. Johnson et al. (2001) found one style, called the English-style hut, that produced low levels of pre-weaning mortality comparable to the farrowing crate in buildings. And more indoor farrowing systems continue to be developed that do not use a restraining crate, often re-inventing elements of the early alternatives.

What the alternatives to the farrowing crate have in common is that they all occupy more floor space than a farrowing crate that might be found in Modern pig farms. Alternatives to the farrowing crates use 9 to 147% more space (average of 48% more) (McGlone, 2002). Floor space in buildings is expensive and so systems that use less floor space and less penning material will be more economical.

In conclusion, farrowing crates do reduce piglet mortality compared with old-style farrowing pens. But reasonable alternatives to the restrictive farrowing crate are available.

2.2 Sow genetics

Pig breeders have been attempting to increase the number of pigs weaned per sow for centuries. In recent decades, pig breeders have understood that the heritability of litter size is reasonably high and would respond to selection pressure (King and Gajic, 1969). So, companies and producers selected for increased numbers of pigs born and weaned. The result is that in the United States, the national average numbers of pigs weaned have risen from 7 to over 10 pigs per sow. All the while, pre-weaning mortality has remained about the same (in the range of 10–15%). Therefore, while we can increase the number of pigs weaned by selecting for litter size, any genetic effect on litter size is probably indirect. For example, sows that sit at all or for longer durations are more likely to crush a pig (moving from the sitting to the lying position). The heritability estimate for sitting is greater than zero and so indirectly one could select against sitting to lower piglet mortality (McGlone et al., 1991). Interestingly, the sloped farrowing pen that reduces piglet mortality also discourages sows from sitting. To improve pre-weaning mortality, environmental conditions and human behaviours need attention.

2.3 Cross-fostering

Piglets are opportunistic sucklers. They will nurse any sow, even if they have a preference for their biological mother. From the sow's perspective, she will accept any piglet that nurses, largely because when she lies down to nurse, she is on her side and cannot push away any piglet that nurses. Cattle and sheep, on the other hand, nurse standing up and they can easily step away if an unwanted neonate attempts to suckle. Therefore, cross-fostering is more easily accomplished in pigs than in ruminants.

Some sows have more piglets than they have functional teats. Domestic pigs of European decent show a behaviour called teat fidelity, which means that a given piglet chooses and defends a single teat. This means that if there are, for example, 12 teats and 13 piglets, then one piglet is sure to die. This extra piglet would be cross-fostered to another sow that had, for example, 11 piglets, but she has 12 functional teats. Cross-fostering can be done in the first few days after birth. The older the piglet, the lower the chance that cross-fostering will be successful.

Low-birth-weight (LBW) piglets are being born today as they have been for decades. The number of piglets per litter is increasing, but the litter weights have not increased (Bovey et al., 2014). The mortality rate for LBW piglets has been reported to be close to 30%. There is a linear increase in the number of LBW piglet mortality and sow parity. In parities 1 through 6 LBW piglets averaged 25.5% mortality versus 9.1% for high-birth-weight (HBW) piglets, but in parities 7 through 12 the mortality rate jumps up to 35.7% in LBW piglets and 11.2% in HBW piglets (Robert et al., 1995). Additionally, processing LBW piglets early in life increases mortality rate. LBW piglets tend to be weaker and many times appear lethargic, when they are processed pain may cause decreased periods of suckling (which has detrimental effect on piglets since they have low energy reserves and immunoglobulins and depend on colostrum for heat production; Herpin et al., 2002), which may lead to death. The decrease in milk consumption may cause piglets to use the udder as a heat source and leads to crushing of piglets; therefore, delaying processing in LBW piglet may improve survival rate (Bovey et al., 2014).

Sows are giving birth to more piglets than they can rear successfully and, therefore, have to be either supplemented or cross-fostered if they are LBW piglets. Many facilities have resorted to transferring piglets to other sows for the purpose of lactating foster piglets, known as nurse sows. The nurse sow is either given newborn piglets once she has finished raising her own or given 1-week-old piglets when her piglets have been weaned (Baxter et al., 2013).

Cross-fostering is stressful for both the piglets and the nurse sows. Piglets cross-fostered after the first day of life have to re-establish a teat order which can cause additional injuries due to fighting (Robert and Marineau, 2001) and disrupted nursings (Heim et al., 2012), which can lead to reduced milk intake and death. Additionally, cross-fostering later than 24 h after farrowing can cause the sows to demonstrate aggressive behaviours towards piglets that are not hers (Pedersen et al., 2008) and can also cause a temporary stoppage of milk let down (Price et al., 1994). Piglets that are cross-fostered the first day after birth are also dirtier and have an increased occurrence in carpal lesions and lameness as a result of fights (Sorensen et al., 2016). On the other hand, nurse sows have a higher tendency to have skin lesions, udder damage and swollen bursae on legs, as a result of prolonged stay in a farrowing crate (Sorensen et al., 2016). Additionally, being a nurse sow takes a toll on the sows' body because they are required to produce milk for a longer period of time, causing body condition score (BCS) to go down because body reserves are depleted throughout lactation (Spencer et al., 2003). Further, low BCS leads to shoulder lesions (Zurbrigg, 2006; Knauer et al., 2007), which can affect sow welfare.

3 Weaning stress

Wild or feral pigs will nurse piglets for three or more months. In commercial pork production pre-industrialized farms (prior to 1960), piglets were traditionally weaned at 8 weeks of age. Creep feed was provided starting at two weeks of age and by the time piglets were 8 weeks of age, they had been gradually weaned. Then, as production efficiencies were calculated, scientists and the industry learned that by weaning 'early', say at 28 days of age, the sow would produce more litters per sow per year and the time to get pigs to market was the same or better. Sow milk production peaks about three weeks after birth, so by four weeks of age, the piglet's appetite is not satisfied by mother's milk alone. The gastrointestinal tract is more developed at four weeks of age than earlier, so weaning is physiologically not problematic.

After the 28-day-of-age weaning phase, economists and pork production specialists concluded that weaning piglets in the 19–23 day range was most profitable. More litters were produced per sow per year, with lower weaning age; however, the piglet's digestive system is less able to handle weaning stress at 19 days of age than 28 days of age. This balance is struck based on economics more than other factors. If one sells pigs into the Modern market, economics may favour weaning at 21 days of age. If one sells in the Natural market, a weaning age of 28 days may be more economically viable because the market itself drives the weaning age higher. Some markets even push for weaning at 35 days of age to make it easier on the piglet to adapt to dry feed. However, one should be aware that with fast-growing, hungry piglets, and high-milking sows, delaying weaning is a metabolic challenge to the sow, with many sows losing significant body weight with advanced weaning age (beyond 21 days). When we observed the behaviour of sows in an outdoor system where sows can walk away from demanding piglets, we saw that sows spend increasing times away from the piglets over time (Johnson et al., 2001). A muscular, demanding 30-day-old piglet can demand significant milk from a lactating sow – and with larger litters, ten or more piglets can nurse with sufficient vigour to cause sows to become very thin. After weaning, sows have an improved metabolic situation with not producing milk, but they still may suffer a form of depression from the loss of piglets in her environment. Much less has been written and investigated about the stress of the sow after weaning. We know that sows experience a significant drop in feed intake and less activity after weaning than before weaning.

The stress that piglets experience at the time of weaning has been widely reported (Niekamp et al., 2007; Moeser et al., 2007; Campbell et al., 2013; Sutherland et al., 2014; Tao et al., 2016) and is known to induce detrimental effects, and sometimes even death. Early weaning adds even more stress because of the piglets' lack of maturity. Weaning may be the most stressful situation a pig experiences, and it is unavoidable in today's Modern production systems.

Prior to weaning, piglets have positive maternal experiences such as maternal milk, warmth, maternal pheromones, interactions with their siblings and other nurturing factors that upon weaning abruptly change (Garcia et. al., 2016). At weaning, pigs are removed from a nurturing environment and must learn to cope with a novel environment in which they must establish a social hierarchy by engaging in fighting, adapt to being without their mother and siblings, a different thermal environment and a different olfactory setting, among other novel/negative experiences. The stress of weaning, specifically the transition between nursing and eating solid foods, may lead to a period of underfeeding that affects growth, metabolism and metabolic changes associated with endocrine adjustments while the animal adapts to the diet change (Le Dividich and Seve, 2000). The disruption of the intestinal barrier (composed of a layer of epithelial cells that line the intestinal tract and serve as a protective line of defence against harmful microorganisms) due to the stress of weaning plays a critical role in disease susceptibility (Spreeuwenberg et al., 2001; Melin et al., 2004; Boudry et al., 2004). Antigenic agents that pass the intestinal barrier lead to inflammation, malabsorption, diarrhoea and potentially systemic disease (Livingston et al., 1995; Deitch et al., 1996; Berkes et al., 2003). Weaning also leads to an abrupt reduction in feed intake, which is extremely low during the first two days after weaning (McCraken et al., 1995) and leads to reduced weight gain, known as 'growth check' or post-weaning 'growth lag' (Tockach et al. 1992; Azian, 1992). The post-weaning growth lag affects economic performance because the time needed to reach market weight is increased (Zijlstra et al., 1996). The acute stress response of weaning can further disrupt homeostasis and compromise well-being in pigs (Niekamp et al., 2007) by increasing cortisol concentrations, altering neutrophil to lymph (N:L) ratio and causing deviations from normal behaviours (Sutherland et al., 2014; McGlone et al., 2016).

3.1 Segregated early weaning

Complete removal of the effects of weaning stress in commercial settings is multifactorial, but we may be able to mediate some of the effects. Segregated early weaning is a common practice to reduce vertical transmission of diseases. It involves removing weaned piglets from the sows and rearing them separately from other age groups, but weaning piglets too early can also have negative effects on growth and well-being. Increasing weaning age from 12 to 21.5 days increases weight sold per pig weaned by 1.80 ± 0.12 kg for each day increased in weaning age, with linear improvements in wean-to-finish growth and productivity that is likely due to a function of both weight and physical maturity at weaning (Main et al., 2004).

3.2 Intermittent suckling

The introduction of intermittent suckling (IS) prior to weaning may help reduce stress at weaning and additionally may help prepare them for the consumption of solid feed before weaning. Providing creep feed to piglets when they are with the sow may not tempt piglets to consume enough due to the satiation of hunger with the sow's milk. However, when piglets are separated from the sow for a certain period (6 to 8 hours), their hunger might be kindled to consume starter feed and thus, prepare them for the change in diet from liquid to solid feed. IS has several benefits, including an increase in eating behaviours shortly after weaning (Berkeveld et al., 2007), increase in average daily gain shortly after weaning (Kuller et al., 2004), less pacing and squealing at weaning that may suggest piglets are less stressed (Brown et al., 2016) and, additionally, IS causes increases in LH secretions (Langendijk et al., 2007), causing sows to ovulate and return to oestrus sooner (Brown et al., 2016). Nonetheless, IS does cause an increase in cortisol when first introduced to piglets and when N:L is increased it did not exhibit any different character than control pigs which are not exposed to IS and weaned at the same time (Turpin et al., 2016). Preparing piglets for weaning with IS is a good alternative to abrupt weaning and has several benefits, such as reducing piglet mortality by allowing the sow to be away from the piglets (possibly reducing avoidance postures and preventing piglet crushing), allowing the sows to recover from the high demand of milk production and increasing the number of litters born each year due to a faster onset of heat.

4 Painful practices: castration and ear notching/tagging

4.1 Physical castration

Piglet processing usually consists of castration, tail docking, corner/ needle teeth clipping or grinding, ear notching, medication injections and identification transponder injections (in some facilities). Among these painful procedures, castration is the most painful in that it causes the most behavioural and physiological changes.

Physical castration is a common procedure performed on young piglets usually at 1-3 days of age and is a standard practice in many countries, including the United States. Physical castration is done to diminish boar taint which causes an unpleasant smell and taste in the meat of sexually mature male pigs and can be aversive to consumers. The undesirable characteristics of boar taint in the meat of uncastrated male pigs are caused by a pheromone known as androstenone and a substance produced in the gut of pigs known as skatole. In order to prevent boar taint when male pigs reach sexual maturity, they are castrated early in life. Castration is typically performed without any type of analgesia (Lumb 2007; Hewson et al., 2007; Rault et al., 2011; Hannson et al., 2011) and, therefore, it is a painful (McGlone and Hellman, 1988; Lumb, 2007; Fredriksen et al., 2011) and stressful procedure in young piglets. Various studies have shown that physical castration leads to changes in behaviour which may be indicative of pain (Wemelsfelder and van Putten, 1985; McGlone and Hellman, 1988; McGlone et al., 1993; Taylor and Weary, 2000; Taylor et al., 2001). Additionally, physical castration can lead to acute changes in physiology including the activation of the hypothalamic-pituitary-adrenal axis and activation of the sympathetic nervous system (White et al., 1995; Prunier et al., 2001). Physiological changes include increased levels of adrenocorticotropic hormone (ACTH), cortisol and lactate after castration, indicative of stress and tissue damage (Prunier et al., 2005).

The use of general and local anaesthetics may be limited by regulations and economics of the swine industry. Most importantly, the use of anaesthetics is not approved in food animals by the FDA in the United States and their use could potentially create problems if these substances are detected in meat. The use of some gaseous anaesthetics may require a gas evacuation system be in place (Prunier et al., 2005), and the use of general anaesthesia has been linked with malignant hyerperthermia in some breeds of pigs (Prunier et al., 2005).

4.2 Alternatives to physical castration

Alternatives to castration include the use of chemical compounds to destroy testicular tissue (Giri et al., 2002; Ljaz et al., 2000), the use of exogenous hormones to downregulate the hypothalamic-pituitary-gonadal axis (Schneider et al., 1998; Daxenberger et al., 2001), immunocastration (IC) (Norman and Litwack, 1997; Hillier, 1994), sperm sexing (von Borell et al., 2009), raising entire males (von Borell et al., 2009), marketing pigs at a younger age and genetic selection against boar taint.

Local destruction of testicular tissue by chemical compounds such as acids and salts may also be painful. Although many authors claim that it castration causes little pain (Prunier et al., 2006), swelling of the testes or scrotum have been observed (Giri et al., 2002), necrosis and slow wound healing (Fordyce et al., 1989).

Downregulation of the hypothalamic-pituitary-gonadal axis by exogenous hormones and IC are two alternatives to physical castration using steroid agonists or antagonists. The latter of the two is known as Improvest® and is used in several countries. The mode of action of IC is through the use of an incomplete version of GnRF (an analogue) in combination with diphtheria toxoid carrier protein for the GnRF analogue (a conjugate). Together, these produce an immunological response in the pig against both the exogenous and endogenous GnRF. The resulting decrease in the levels of FSH and LH further suppresses reproductive function (Esbenshade and Britt, 1985). Without LH and FSH stimuli the gonads will not develop and the animal will stay in a pre-pubertal condition, or the gonads will atrophy in the post-pubertal animal. Currently, Improvest® is approved in more than 60 countries, but there is no market for it in the United States due to low acceptance by consumers. Other problems with its use include laborious administration, high cost and the risk of self-injection in humans (Prunier et al., 2006).

There are several alternatives to castration, but so far none has really been permanently adopted, and physical castration continues to be the more prevalent method of castration in the United States and in some other countries. Therefore, pain during and after physical castration still remains a valid concern. The use of analgesia and/or local anaesthesia to physically castrate piglets has been adopted by several countries and may be one of the most viable ways to reduce pain during and after castration until better alternatives are found and adopted worldwide.

4.3 Ear notching/ear tags

The need for identification of individual animals is a key point for management, traceability, trade control and disease eradication (Caja et al., 2005), but has to be individualized, permanent, simple to apply and read, welfare appropriate and tamper-proof (Merks et al., 1990; McKean, 2001). The use of ear notching for identification has greatly decreased but was in the past the most common method to standardize purebred and commercial swine (Rea, 1986). It was typically performed by cutting deep notches into the ears approximately 6.5 mm deep (Widoski and Torry, 2002). Some small producers may still use ear notching as a means of identifying purebred pigs. Little research has been conducted to assess the degree of pain and distress associated with ear notching (Widoski and Torrey, 2002), or even comparing ear notching to ear tagging (Marchant-Forde et al., 2009). However, ear notching can cause pain and irritation, indicated by head shaking and scratching of the ears with hind limbs (Noonan et al., 1994). Marchant-Forde et al. (2009) reported that ear-notched pigs have worse wound scores and higher cortisol concentrations than ear-tagged pigs, and emit calls higher in frequency than control pigs. Most recently, the use of ear tags has supplanted the use of ear notching, but little is known about the pain caused due to ear tagging.

4.4 Alternatives to ear notching/ear tags

Identifying individual animals is a must and therefore, the need for an efficient method of identification led to the development of radio frequency identification (RFID). RFIDs can be placed either under the skin as ear tags (usually injected in the auricle base of the ear base) or intraperitoneally (IP). There are three main types of tags used for animal identification – boluses, ear tags and injectable glass tags – and can be grouped into low or high frequency (Voulodimos et al., 2010). Although RFIDs are probably less painful than ear notching and more efficient (because improperly notched pigs lead to animal misidentification) they still have pit falls. Conventional ear tags can be soiled with faeces, can fade and be hard to read, can get pulled off or chewed on and can also lead to misidentification of pigs; however, RFID ear tags have very similar problems. Firstly, the person trying to identify the animal must be close enough to scan the tag and if it is damaged, identifying the animal can be a problem. Out of the three placements for RFIDs, intraperitonally injected transponders were not problematic and are an efficient method for traceability of pigs (Caja et al., 2005). Intaperitonally placed RFIDs require injection of the transponder into the abdominal cavity between the intestines, a task that requires trained personnel (as the injection must be performed on the left, ventral side of the animal, at ~1 cm from the ventral line and 2cm caudal to the navel) because misplacement into the intestines and urinary bladder is possible (Caja et al., 2005). However, Babot et al. (2006) found that losses of IP transponders were very low. They further reported that on-farm losses averaged 1.6% for ear tags and1.8% for IP injected transponders, but electronic failures seem to occur in ear tags and not IP injected transponders. Thus, Babot et al. (2006) suggested that IP injected transponders give the best results compared to SQ injections and electronic ear tags (damage to the tags and losses during transport to slaughter). An additional advantage of RFIDs is that they offer automatic reading in dynamic conditions, but they cost 7 to 10 times more than conventional ear tags and must include the purchase of a reading device (Babot et al., 2010). Furthermore, electronic devices must also remain functional at the slaughter line, and removal of these devices from the carcass within 5 seconds after slaughter is considered acceptable – a task which may be difficult due to slaughtering conditions of pigs that may include using fire, hot water and a high line speed (Merks and Lambooy, 1990).

Although there are not many reports in the literature referring to negative effects on animal welfare due to the use of RFIDs, Caja et al. (2005) and Babot et al. (2006) both agree that these do not have a negative effect either on animal welfare (as no inflammatory reactions were observed and animals healed quickly) or on performance.

Other novel methods of identification include the use of auricular vein patterns (Wayne et al., unpublished), biometrics such as retinal images (Rusk et al., 2006) and facial recognition (Corkery et al., 2007). Harrel and colleagues used an OptiReader™ and Veinlite™ LED to image auricular veins in pigs and found that each pig has a distinct auricular vein pattern. They imaged vein patterns of 96 pigs, resulting in a match rate of 96.7% with an imaging error of 8.3% (mainly due to the colour of the pig, black ears were harder to match). They concluded that further research is needed to determine the uniqueness of each pig's auricular vein pattern and the proper distance for imaging. The use of retinal images is a technology that has been reported to be a viable method of identification for beef and sheep (Rusk et al., 2006; Allen et al., 2008; Adell et al., 2012). Facial recognition is a common biometric used in humans and is based on location and shape of facial attributes and overall analysis of the face image (Corkery et al., 2007). Corkery et al. (2007) found that the recognition rate for sheep was 95.3% to 96%, and is non-invasive and inexpensive.

5 Painful practices: tail docking and teeth clipping/resection

5.1 Tail docking

Piglets can many times become victims of tail biting, and therefore this is the main reason that tail docking is an accepted procedure in the swine industry. However, tail docking is painful and the use of anaesthetics and analgesics is not a common practice to reduce pain during or after tail docking. It is important to understand that tail docking reduces the incidence of tail biting, but does not eliminate tail biting entirely (Paoli et al., 2016). Harley et al. (2014) reported that out of 99% of pigs' tail docked in a study conducted in Ireland, 72.5% still presented mild tail-bite lesions at the abattoir and 2.5% of those were severe lesions.

Tail biting can result in infections and severe injuries (causing acute pain) to the tail, such as tails being bitten to the rump (leading to the victim having to be euthanized; Sutherland and Tucker, 2011). Additionally, severe tail biting can lead to reduced pig performance (van Staavern et al., 2016) and also lead to financial losses after meat inspection at the abattoir as a result of carcass condemnations due to lesions associated with tail biting (Harley et al., 2014).

Tail biting decreases by increasing docking length (Thodberg et al., 2010), but, in general, about 1.5 to 2 cm is left of pigtails (Hunter et al., 2001; Sutherland et al., 2011). Tail docking is usually performed with cutting pliers, knives, scalpel blades or hot docking irons. Although there is a benefit to tail docking due to the reduced incidence of tail biting, it is still considered a painful procedure. Tail docking can cause both acute and possibly chronic stress (Herskin et al., 2015) in addition to physiological and behavioural changes in piglets (Sutherland and Tucker, 2011). Prunier et al. (2005) reported that adrenocorticotropin hormone (ACTH) and lactate concentrations do not differ between tail-docked pigs and control pigs at one day of age. The cortisol response also does not differ between docked and control pigs at 1 or 6 days of age when a cautery iron is used (Prunier et al., 2005; Sutherland et al., 2008). Yet Sutherland et al. (2008) reported higher cortisol concentrations at 60 minutes and at six days of age when pig tails were docked with cutting pliers compared to control pigs. Hot iron cautery docking is indicated to be less stressful compared with docking by cutting off the tail (Sutherland et al., 2008), but can lead to slower healing (Sutherland et al., 2009) and development of neuroanatomical changes, such as neuromas in the outermost tail tip (Herskin et al., 2015; Sandercock et al., 2016).

Pigs have a need for manipulable materials to satisfy a range of behavioural needs (intrinsically motivated exploratory and foraging behaviour) and if these are not met there is an increased incidence of tail-biting behaviours, especially in weaner and rearing pigs (EFSA, 2014). Tail biting outbreaks in swine facilities can be multifactorial (Sutherland et al., 2008), including lack of manipulable material, poor climate, poor air quality, poor health, nutritional deficiencies, competition for resources, social instability, high stocking density, poor pen layout and high genetic potential for lean tissue growth (Fig. 1; EFSA, 2014).

5.2 Alternatives to tail docking

Managing tail biting is complex and many factors have to be taken into consideration. All farms are different and one thing may work on one farm and not in another; therefore, adhering to what is effective by site may be one of the most effective ways to reduce tail biting.

RISK FACTORS (resources, environmental and management factors)	WELFARE CONSEQUENCE (tail-biting only considered)	ANIMAL-BASED INDICATORS (in relation to the welfare consequence)
Lack of manipulable material (quantity or quality, withdrawal, or absence in early life)		Occurrence of bitten tails
Poor climate (too hot/cold, draughts)		Tail manipulation behaviour increased
Poor air quality	Tail-biting associated detrimental welfare consequences :	Tail length shortened
Poor health	- pain - fear	Increased presence of abscesses at abattoir level
Nutritional deficiency (minerals, amino acids)	- infection - disturbed rest	Tail posture lowered
Competition for resources (feeding, drinking, lying)	- disturbed feeding - altered behaviour	
Social instability (mixing, moving)		
High stocking density		
Poor pen layout (disturbed resting; poor pen hygiene)		
High genetic potential for lean tissue growth		

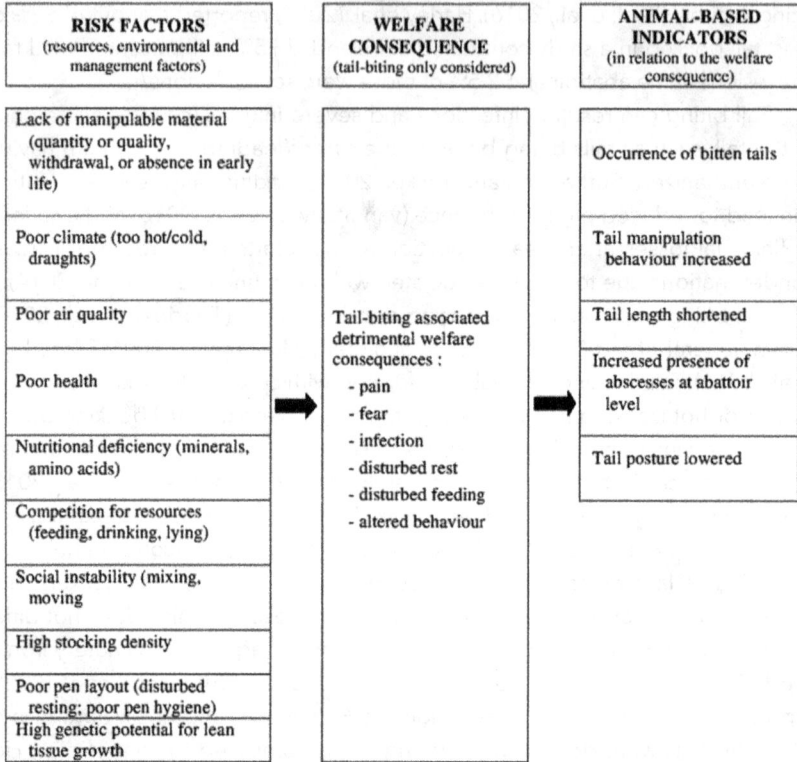

Figure 1 Associations between risk factors, welfare consequence and animal-based indicators of tail biting (risk factors to be controlled to avoid tail biting). Adapted from EFSA, 2014.

One of the alternatives commonly suggested is the use of manipulable materials, such as straw hay. Straw hay has been reported to reduce tail biting by 50% (Hunter et al., 2001; Zonderland et al., 2008). Other enrichment objects have been studied but pigs seem to prefer objects that are ingestible, have chewable properties and destructibility (Van de Weerd et al., 2003).Pigs lose interest rather quickly when provided with toys compared to the use of manipulating straw, as they spent 2% of their time manipulating a toy compared to 20% of their time manipulating straw (Scott et al., 2006). Although straw is a good alternative because it allows rooting behaviours that distracts pigs, which shifts them away from tail biting, it is difficult to dispose of, especially with manure systems that cannot handle this type of material.

Another alternative is the removal of the tail biter. Removing the tail biter fixes the problem temporarily but in the experience of the author, it seems to not always be just one culprit, but several. Identifying the culprit(s) is time consuming and finding space for biters in barns can also be difficult in small to medium farms. Therefore, a combination of risk factors must be taken into consideration and the option that best works for each specific farm should be used.

The use of pig advisory services improves farm productivity through better housing, management and nutrition, and positively affects pig welfare by promoting detailed record keeping of financial and farm performance records (van Staaveren et al., 2016). The Teagasc Advisory Service provides farms with consultation visits, herd performance and financial monitoring, organizes discussion groups and provides education and training for the pig sector (van Staaveren et al., 2016). van Staaveren et al. (2016) reported that keeping of financial and performance-related records with the Teagasc eProfit Manager (ePM; which monitors over 150 parameters detailing technical and performance data of the herd) system was associated with a lower prevalence of moderate tail lesions, and further suggests that improving performance is likely associated with improvements in welfare.

5.3 Teeth clipping/resection

Very few piglets on US farms are now teeth clipped. While it was a common practice before in the 1990s, farms discovered that teeth clipping of all the piglets was not necessary. Today, skilled workers target teeth clipping to the piglets who scar other piglets or the sow's udder. We estimate that fewer than 10% of piglets on Modern US farms have their teeth clipped today.

Needle teeth, the deciduous canines and the corner incisors (Hay et al., 2004) are sharp and pointed, protrude out from the jaw at an angle and, therefore, can cause damage to other piglets and sow udders. Piglets display aggressive behaviours towards siblings to establish teat order and cause lesions to the faces of littermates (Marchant-Forde, 2009). Although resection has proven to reduce face injuries in piglets, it may also cause internal mouth lesions for piglets with resected teeth (Lewis et al., 2005a), resulting in pulpits (Hay et al., 2004), deep wounds to the tongue and lips (Burger, 1983), and splinters that can become embedded in the gums (Hunter et al., 1993). Additionally, the wounds seen with teeth resection may be the entry points for infection (Hunter et al., 1993) and could possibly lead to death. Hay et al. (2004) found that clipping teeth led to higher numbers of teeth with opening of the pulp cavity, teeth with fractures, haemorrhage, infiltration and abscesses compared to intact teeth, and resection was painful (with pain lasting up to the 15th day of life). Teeth-clipped pigs display irregular behaviours, such as chomping (Lewis et al., 2005a), but there are very few reports on the effect teeth clipping has on piglets. Chomping is attributed to irritation, pain or discomfort due to possible presence of teeth fragments and blood in the mouth (Batailee et al., 2002).

5.4 Alternatives to teeth clipping

One of the alternatives to replace teeth clipping is grinding of the teeth with a rotating grindstone. However, teeth grinding increases the mean call vocalizations, negatively affects body weight, growth rates tend to be lower, cortisol concentrations increase compared to clipping of the teeth one week post-treatment, and takes longer to perform than clipping of the teeth (Merchant-Forde et al., 2009). Additionally, teeth grinding may possibly cause additional handling stress for the pigs. Lewis et al. (2005a) suggested that grinding the teeth has some negative piglet welfare implications, but is better than the pain and discomfort associated with teeth clipping. However, they found that teeth clipping was more effective at reducing facial injuries in piglets than grinding. They concluded that grinding is recommended over clipping or leaving the teeth intact.

EU legislation permits both teeth clipping and grinding (Directive, 2001/93/EC), but only when there is evidence that there are injuries due to piglet aggression and injury to sow teats. Even when there is evidence, producers must have altered the environment and management practices in some way to prevent aggressive behaviours (Lewis et al., 2005a,b). This philosophy is largely followed in the United States.

The other alternative to teeth clipping is to leave teeth intact only in LBW piglets. There are two types of sibling competition: (1) direct competition – aggressive competition for teats, mainly in the first hours after birth, when ownership of a teat is established and where litter mates die or survive by suckling opportunistically (Hartsock and Graves, 1976; de Passille et al., 1988, 1989) – and (2) indirect competition – larger or more competitive piglets gain more weight than their litter mates, by possibly stimulating and draining their teats more efficiently and garnering to their respective teats more hormones and nutrients (Fraser et al., 1979; Thompson and Fraser, 1986). LBW piglets are disadvantaged in both types of competition (fail to secure their own teat and die during the first few days of life) and, therefore, by selectively tooth clipping only the larger piglets in the litter, LBW piglets have a decrease in mortality (as the presence of their needle teeth may allow them to defend their teat better) and weight gain is greater (Robert et al., 1995). Robert et al. (1995) found that in larger litters (12–14 piglets) LBW piglets that did not have their teeth clipped had a lower mortality rate than control litters which did have their teeth clipped (32% mortality compared to 39%, respectively), whereas HBW piglets had a trend in the opposite direction (14.4% mortality in the experimental groups and 13.2% in the controls). They concluded that selective tooth clipping works by assisting small litter mates (as intact teeth contribute to indirect competition) at the expense of the larger ones (only in large litters) contributes to uniform weaning weights and helps the most vulnerable pigs stay alive until fostering or another intervention, but does not improve overall growth and survival.

Leaving needle teeth intact in all piglets may not be a logical alternative since it can also lead to increased piglet mortality and damage sow udders. Farrowing crate restrictions prevent sows from expressing normal behaviours that would otherwise be expressed in the wild. In semi-natural conditions, sows remain in the nest with piglets for the first few days after birth (Lewis et al., 2005b), but after the first week they begin to leave the nest for long periods of time, returning only to suckle their piglets (Jensen and Rhebo, 1987). In farrowing crates, sows cannot reduce the amount of contact with their piglets and piglet behaviour directed towards the udder is very high (Cox and Cooper, 2001), and thus, sows begin to change postures more frequently (more ventral lying and dog sitting during lactation; Cronin et al., 1992, 1998). The change in postures, possibly due to frustration, leads to accidental crushing of piglets. Lewis et al. (2005b) concluded that intact teeth injured sows and caused discomfort and linked sow avoidance behaviour to increased piglet mortality. They further reported that teeth clipping was more beneficial to sow welfare than grinding or intact teeth due to the number of teat lesions observed on sows being less (at 11 days post-partum) when piglets were teeth clipped.

6 Transportation

Piglets are often transported to other facilities upon weaning, which causes additional stress to piglets. Currently in the United States, animals can be transported for 28 h without provision of water and feed or off-loading. It was previously believed that weaning stress was so great that transportation did not have an effect. Piglets are many times transported for many hours without feed and water. When piglets are transported for more than 24 h without feed and water, there is an apparent loss in body weight, negative effects on physiology and behaviour (Garcia et al., 2015). Garcia et al. (2016) reported that weaned pigs not provided with water during 32 h of transport lost significantly more weight than those that did have access to water, and further suggested that provision of water was more important than feed during transportation. Regardless of provision of feed and water, transportation is a complex stressor made up of many factors including fluctuating temperatures, stocking density, mixing with unfamiliar pigs and motion (Lambooij et al., 1993). Therefore, transportation has the potential to affect the health and welfare of pigs, especially in pigs already experiencing weaning stress; therefore, weaning stress and transport can cause additive negative effects. Transport of weaned pigs has many common factors with transport of finishing pigs, including mixing, fasting, temperature fluctuations, vibration, novelty and noise (Lewis et al., 2008). Similar to studies in finishing pigs, transport has been shown to cause behavioural and physiological changes indicative of stress in weaned pigs (Sutherland et al., 2014). Finding methods to reduce stress during transport could help reduce some of the negative effects on piglet performance after arrival to new facilities.

7 New technologies

Piglets have few biological products that are commonly used. These include supplemental iron, antibiotics and vaccines. Some nutritional supplements may have value. There are opportunities for innovative approaches to improve piglet health and welfare that are from entirely new classes of biological or electronic technologies.

7.1 Pheromones, interomones and semiochemicals

Pigs have a large number of olfactory receptors – four times as many as humans (Nguyen et al., 2012). Pigs are considered macro-osmotic while humans are micro-osmotic species. In addition, the pig's olfactory acuity is thousands-fold more sensitive than humans; that is, they can detect odour molecules at a much lower concentration. Impacting the olfactory environment may hold promise to improve pig health and welfare.

One can think of semiochemicals (pheromones, interomones and inter-species biologically relevant odours) as we do nutrients. When a nutrient is deficient, pigs grow slower and have health problems. When a pig is weaned from its mother, it is missing the maternal and some adult male odours from its environment. A lack of maternal-neonatal-rich olfactory environment contributes to the stressfulness of weaning. Alarm pheromones contribute to stress at other times. A lack of adult odours can encourage fighting among neonates. Unfamiliar neonates or sows must fight until they understand each other's olfactory signals and ultimately its place in the social order.

Pheromones are an important releaser of a variety of behaviours in the receiving animal, and can be found in faeces, urine, saliva, vaginal mucus and/or secreted from cutaneous scent glands (Rekwot et al., 2001; Mucignat-Caretta, 2014). They are classified as molecules released by individuals that elicit a behaviour or physiological change in the endocrine or reproductive system of other individuals of the same species (Tirindelli et al., 2009). Although pheromones typically act among individuals of the same species (intraspecific), they can be detected by different species (interspecific). An interomone is a pheromone in one species that changes the behaviour of another species without the requirements to benefit or harm the emitter or the receiver (McGlone et al., 2014).

The use of pheromones is not new to pork production. The boar salivary pheromone, androstenone, was reported to reduce aggressive behaviour in piglets in the 1980s (McGlone, 1985; McGlone and Morrow, 1988). Androstenone also reduces unwanted behaviours in other species of animals (acting as an interomone), such as barking and jumping in dogs (Stop That! Sentry, Omaha, NE) (McGlone et al., 2014).

Morrow and McGlone (1990) described sources of maternal odours in lactating sows. They showed that skin washings from sows are recognized by the piglet starting at a few hours of age. Patrick Pageat and his research group collected similar biological fluids and called them pig appeasing pheromones (Pageat, 2001; US patent 6,077,867). McGlone and Anderson (2002) showed that this mixture of fatty acids found on a sow's udder will stimulate weight gain, reduce fighting and increase feeding behaviours in weaned pigs.

A second maternal-neonatal odorous molecule was then discovered to be biologically active in the pig. The interomone 2-methyl-2-butenal (2M2B) is a rabbit maternal pheromone that elicits nursing behaviour in rabbit pups. Recently, 2M2B was shown to be beneficial in pigs by improving post-weaning performance (McGlone et al., 2016). Interestingly, 2M2B increased growth in newly weaned piglets (McGlone et al., 2016), but has also been marketed as a calming spray or collar for cats and dogs (Perrigo Animal Health). Animals spend the majority of their life communicating through olfactory signals. One way of improving animal experiences in a production setting is to incorporate the use of volatile compounds or semiochemicals such as pheromones into their environment. Animals, especially pigs 'see' their world through olfaction, due to the high number of olfactory receptors (Groenen et al., 2012). Therefore, manipulating the olfactory environment in pig production settings may provide numerous benefits.

We expect more semiochemicals will be discovered in pigs and other farm animal species. Modulation of the olfactory environment can improve reproduction, reduce stress and improve performance using what has been one approach that has been called 'clean, green and ethical' technology (Scaramuzzi and Martin, 2008).

7.2 Automated detection systems

Our pig barns are far from smart barns. While we may carry a phone that is more powerful than a mainframe computer of 40 years ago, the pig barn and pig pasture are simple systems. They have not evolved as have consumer, military or business electronics. However, the pig production unit will evolve as technology evolves.

A smart barn would have many automated features. Labour is hard to find in commercial agriculture. Automated systems could replace some people in production agriculture. Furthermore, automated systems would potentially do a better job at 24/7/365 surveillance of a pig barn or pasture. Scientists and engineers are developing sensors such as cough detectors, piglet scream sensors (for when a sow starts to crush a piglet), pig counting and body weight sensing devices, among others. Livestock and poultry automation in the future is as sure as tractors replacing humans and draft horses and milking machines replacing human milkers.

Table 1 Current levels of automation for each behavioural category. Adapted from Matthews et al., 2016

Behavioural categories	Automation categories		
	Behaviour detection	Behaviour monitoring	Automated detection of behavioural change
Daily activity budget	Requires monitoring behaviour over time	Location-based (Andersen et al., 2014); Locomotor activity (Chung et al., 2014); Drinking (Madsen and Kristensen, 2005).	
Feeding, drinking, and elimination	Drinking (Meiszberg et al., 2009).	Feeding (Fernández et al., 2011; Andersen et al., 2014; Maselyne et al., 2015b); Commercial systems: I-BOX 360°, Farmex, Eliskool 2/Elister 2/Tristar systems, IVOG, FIRE; Elimination (Zhu et al., 2009).	Drinking (Madsen and KtifletKto.HK)
Posture and locomotion	Locomotion (Lind et al., 2005; Kongsro, 2013).	Spatial distribution (Cook et al., 2015; Nasirahmadi et al., 2015); gait (Starvrakakis et al., 2014); General activity (Leroy et al., 2006).	Activity (Martínez-Avilés et al., 2015).
Social behaviour	Aggression (Oczak et al., 2012, 2014; Viazzi et al., 2014; Lee et al., 2016); Clustening (Shao and Kin, 2008); Vocalisation (Maniteuffel and Schön, 2002; Schön et al., 2004).		
Disease-specific	Coughing (Chedad et al., 2001; Exadaktylos et al., 2008; Chung et al., 2013).	Coughing (Vandermeulen et al., 2013; Hemeryck and Berckmans, 2015; Hemeryck et al., 2015).	

Health and welfare compromises in pigs have wide-ranged consequences. Automatic detection of compromised health and welfare can be accomplished with new technology that includes measures of health, growth and behaviour (Matthews et al., 2016). Audio, visual and other modalities that automatically monitor behaviours are certain to be of great benefit to the pig industry. We show some reported applications of this technology below (Table 1).

Audio sensors (automated processing techniques) can be used to detect and classify certain acoustic events such as coughing, sneezing, screaming, barking and the assessment of this audio data (through spectral analysis) can help classify sick pigs (Matthews et al., 2016). An example of an audio system is the STREMDO, a system that automatically measures the duration and intensity from vocalizations (Manteuffel and Schon, 2002; Schon et al., 2004). Audio sensing can also be used to measure external stimuli coming from the environment that may have behavioural impacts on the animals (Marx et al., 2003; Broucek, 2014).

Video is another automated system that can be used to measure general pig activity. Pixel differences between consecutive images (provide a measure of the group rather than an individual; Rushen et al., 2012) can detect fast pig movements that may be indicative of aggression, but can also be due to other behaviours such as playing and chasing (Viazzi et al., 2014). Non-invasive measurements (that do not require handling) of surface temperature in pigs from infrared video (McManus et al., 2016) have supported the evaluation of automated assessment of thermal comfort measured from thermal distribution of pigs (Cook et al., 2015).

The use of sensors to measure water consumption (Meiszberg et al., 2009), RFID at feeding and drinking (Andersen et al., 2014; Maselyne et al., 2015a,b), accelerometers combined with temperature sensors to detect abnormal body temperatures due to infection (Martinez-Aviles et al., 2015) and pressure mats that measure animal gaits for detection of lamenesses (Meijer et al., 2014) are all modalities that may aid in reducing the amount of labour and cost associated with labour, along with improving animal welfare.

7.3 Gene editing

Animal health challenges will be one of the biggest challenges for efficient livestock production in the future, due to resistance to existing drugs and the pressure to reduce antibiotic use in agriculture (Plastow, 2016). According to Plastow (2016), genome-wide analysis studies identify genetic factors influencing variation in how animals respond to diseases (genetic variation in the susceptibility to disease), which can be polygenic or even oligogenic (depending on the number of genes distributed across the genome). Boddicker et al. (2014) found that for porcine reproductive respiratory syndrome virus (PRRSV) one region on chromosome 4 was found to explain more than 10% of the variation in viraemia and growth after infection and reported the effect to be dominant so that only one copy of the allele is necessary to get the benefit. Further, Hess et al. (2014) confirmed the previous findings using a second strain of the virus and suggested that it could be possible to select for pigs that are less impacted by PRRSV when it is present. Scientists have developed genetically modified (GM) pigs intended for commercial production: One has a gene inserted to increase milk production and another has an enzyme added to break down plant phytase to make phosphorous more available (which has a large environmental advantage). Both GM pigs did not enter commercial herds largely due to perceived consumer negative reactions.

Gene editing is a newer technique that is being investigated (Perez-Pinera et al., 2012). In gene editing, the natural genome is changed to activate or inactivate genes. One viable example is a pig that had a PRRSV receptor protein inactivated. These pigs do not succumb to PRRSV infection. One can see the economic and welfare value to such a gene-edited pig. The consumer reaction to such a gene-edited pig is yet to be seen.

Gene editing may hold great potential as a way to improve animal welfare. The disease-resistant opportunities are as many as we have infectious diseases. However, beyond infectious diseases, one can imagine many other welfare advantages. Imagine a pig born with no needle teeth to clip, or male pigs that do not need to be castrated because the boar taint molecules are suppressed. For every animal welfare and animal health problem, there is a potential gene editing opportunity. But ethical thinking and consumer understanding are far behind in knowing how to handle these opportunities.

8 Conclusion

The commercial swine industry has changed both the production environment and the genetics of the pig significantly in the past half century. At the same time that pigs grow faster, are leaner and have larger litters, the consumers have adopted a variable view of production agriculture. A half century ago, low-cost food was the primary consideration for both producer and consumer. Today, we have a mixed bag of consumer desires. Some consumers still seek low-cost, high-quality pork. Others care about the environment, food safety, taste and animal welfare, among other concerns. Social media has accelerated the plethora of views among consumers. Bloggers and videographers have a powerful effect on consumer trends. At the same time, retail food companies (restaurants and grocery stores) have requirements for on-farm practices including animal welfare. All parties share the goal of improving animal welfare. The pork industry has opportunities to improve animal welfare and at the same time profitability and sustainability of the agricultural enterprise. The four major areas to improve piglet welfare are to reduce pre-weaning mortality, reduce weaning stress, prevent pain during standard practices and reduce stress associated with the transport of piglets. But in addition to making progress in these four long-investigated areas, we have opportunities to improve pig welfare in new areas such as pheromones, automated technologies and gene editing.

9 Where to look for further information

For further information about these topics or other animal well-being topics, please refer to Texas Tech University Laboratory of Animal Behavior, Physiology and Welfare (www.depts.ttu.edu/animalwelfare/).

10 References

Adell, N., Puig, P., Rojas-Olivares, A., Caja, G., Carné, S. and Salama, A. A. (2012). A bivariate model for retinal image identification in lambs. *Computers and Electronics in Agriculture*, 87, 108–112.

Allen, A., Golden, B., Taylor, M., Patterson, D., Henriksen, D. and Skuce, R. (2008). Evaluation of retinal imaging technology for the biometric identification of bovine animals in Northern Ireland. *Livestock Science*, 116(1), 42–52.

Andersen, H. L., Dybkjær, L. and Herskin, M. S. (2014). Growing pigs' drinking behaviour: number of visits, duration, water intake and diurnal variation. *Animal*, 8(11), 1881–8.

Babot, D., Hernández-Jover, M., Caja, G., Santamarina, C. and Ghirardi, J. J. (2006). Comparison of visual and electronic identification devices in pigs: On-farm performances. *Journal of Animal Science*, 84(9), 2575–81.

Bataillie, G., Rugraff, Y., Chevillon, P. and Meunier-Saluaum, M. (2002). Caudectomie et section des dents chez le porcelet: conséquences comportementales, zootechniques et sanitaires. *Techni Porc*, 25(1), 5–13.

Baxter, E. M., Rutherford, K. M.D., D'Eath, R. B., Arnott, G., Turner, S. P., Sandøe, P., Moustsen, V. A., Thorup, F., Edwards, S. A. and Lawrence, A. B. (2013). The welfare implications of large litter size in the domestic pig II: management factors. *Animal Welfare*, 22, 219–38.

Berkes, J., Viswanathan, V. K., Savkovic, S. D. and Hecht, G. (2003). Intestinal epithelial responses to enteric pathogens: effects on the tight junction barrier, ion transport, and inflammation. *Gut*, 52(3), 439–51.

Berkeveld, M., Langendijk, P., Bolhuis, J. E., Koets, A. P., Verheijden, J. H. and Taverne, M. A. (2007). Intermittent suckling during an extended lactation period: Effects on piglet behavior. *Journal of Animal Science*, 85, 3415–24.

Boddicker, N. J., Bjorkquist, A., Rowland, R. R., Lunney, J. K., Reecy, J. M. and Dekkers, J. C. (2014). Genome-wide association and genomic prediction for host response to porcine reproductive and respiratory syndrome virus infection. *Genetics Selection Evolution*, 46, 1.

Boudry, G., Péron, V., Le Huërou-Luron, I., Lallès, J. P. and Sève, B. (2004). Weaning induces both transient and long-lasting modifications of absorptive, secretory, and barrier properties of piglet intestine. *The Journal of Nutrition*, 134(9), 2256–62.

Bovey, K. E., Widowski, T. M., Dewey, C. E., Devillers, N., Farmer, C., Lessard, M. and Torrey, S. (2014). The effect of birth weight and age at tail docking and ear notching on the behavioral and physiological responses of piglets. *Journal of Animal Science*, 92(4), 1718–27.

Brouček, J. (2014). Effect of Noise on Performance, Stress, and Behaviour of Animals. *Slovak Journal of Animal Science*, 111.

Brown, E. G., Krebs, L. B., Boone, C. L. and Cauthen, T. (2016). Effects of Intermittent Suckling on Sow and Piglet Performance. *Texas Journal of Agriculture and Natural Resources*, 22, 55–60.

Burger, A. (1983). Consequences of clipping incisors in piglets (Doctoral dissertation, Thesis, Tierarztliche Fakultat Ladwig-Maximilians-Universitat, Munchen, Deutschland).

Caja, G., Hernandez-Jover, M., Conill, C., Garin, D., Alabern, X., Farriol, B. and Ghirardi, J. (2005). Use of ear tags and injectable transponders for the identification and traceability of pigs from birth to the end of the slaughter line. *Journal of Animal Science*, 83, 2215–24.

Campbell, J. M., Crenshaw, J. D. and Polo, J. (2013). The biological stress of early weaned piglets. *Journal of Animal Science and Biotechnology*, 4(1), 1.

Cook, N. J., Chabot, B., Lui, T., Bench, C. J. and Schaefer, A. L. (2015). Infrared thermography detects febrile and behavioural responses to vaccination of weaned piglets. *Animal*, 9(02), 339–46.

Corkery, G. P., Gonzales-Barron, U. A., Butler, F., Mc Donnell, K. and Ward, S. (2007). A preliminary investigation on face recognition as a biometric identifier of sheep. *Transactions of the ASABE*, 50(1), 313–20.

Cox, L. N., and Cooper, J. J. (2001). Observations on the pre-and post-weaning behaviour of piglets reared in commercial indoor and outdoor environments. *Animal Science*, 72(1), 75–86.

Cronin, G. M., Dunsmore, B. and Leeson, E. (1998). The effects of farrowing nest size and width on sow and piglet behaviour and piglet survival. *Applied Animal Behaviour Science*, 60(4), 331–45.

Cronin, G. M., Simpson, G. J. and Hemsworth, P. H. (1996). The effects of the gestation and farrowing environments on sow and piglet behaviour and piglet survival and growth in early lactation. *Applied Animal Behaviour Science*, 46(3), 175–92.

Daxenberger, A., Hageleit, M., Kraetzl, W. D., Lange, I. G., Claus, R., Le Bizec, B. and Meyer, H. H. (2001). Suppression of androstenone in entire male pigs by anabolic preparations. *Livestock Production Science*, 69(2), 139–44.

De Passille, A. M. B. and Rushen, J. (1989). Suckling and teat disputes by neonatal piglets. *Applied Animal Behaviour Science*, 22(1), 23–38.

De Passille, A. M. B., Rushen, J. and Hartsock, T. G. (1988). Ontogeny of teat fidelity in pigs and its relation to competition at suckling. *Canadian Journal of Animal Science*, 68(2), 325–38.

Deitch, E. A., Rutan, R. and Waymack, J. P. (1996). Trauma, shock, and gut translocation. *New Horizons*, 4(2), 289–99.

EFSA Panel on Animal Health and Welfare 2007. Scientific Opinion of the Panel on Animal Health and Welfare on a request from Commission on the risks associated with tail biting in pigs and possible means to reduce the need for tail docking considering the different housing and husbandry systems. *EFSA Journal* 611, 1–13.

Eide, I. (1963). Farrowing crate. USA patent 3,077,861. Publication date 19, Feb 1963.

Esbenshade, K. L. and Britt, J. H. (1985). Active immunization of gilts against gonadotropin-releasing hormone: effects on secretion of gonadotropins, reproductive function, and responses to agonists of gonadotropin-releasing hormone. *Biology of Reproduction* 33(3), 569–77.

Fahmy, M. H. and C. Bernard. (1971). Causes of mortality in Yorkshire pigs from birth to 20 weeks of age. *Canadian Journal of Animal Science*, 51, 351–9.

Fordyce, G., Beaman, N., Laing, A., Hodge, P., Campero, C. and Shepherd, R. K. (1989). An evaluation of calf castration by intra-testicular injection of a lactic acid solution. *Australian Veterinary Journal*, 66(9), 272–6.

Fraser, D. (2003). Assessing animal welfare at the farm and group level: the interplay of science and values. *Animal Welfare*, 12(4), 433–43.

Fraser, D., Thompson, B. K., Ferguson, D. K. and Darroch, R. L. (1979). The 'teat order' of suckling pigs: 3. Relation to competition within litters. *The Journal of Agricultural Science*, 92(02), 257–61.

Fredriksen, B., Johnsen, A. M. S. and Skuterud, E. (2011). Consumer attitudes towards castration of piglets and alternatives to surgical castration. *Research in Veterinary Science*, 90(2), 352-7.

Garcia, A., Pirner, G., Picinin, G., May, M., Guay, K., Backus, B., Sutherland, M. and McGlone, J. (2015) Effect of provision of feed and water during transport on the welfare of weaned pigs. *Animals*, 5(2), 407-25.

Garcia, A., Sutherland, M., Pirner, G., Picinin, G., May, M., Backus, B. and McGlone, J. (2016). Impact of Providing Feed and/or Water on Performance, Physiology, and Behavior of Weaned Pigs during a 32-h Transport. *Animals*, 6(5), 31.

Giri, S. C., Yadav, B. P. S. and Panda, S. K. (2002). Chemical castration in pigs. *Indian Journal of Animal Sciences*, 72(6), 451-3.

Groenen, M. A., Archibald, A. L., Uenishi, H., Tuggle, C. K., Takeuchi, Y., Rothschild, M. F., Rogel-Gaillard, C., Park, C., Milan, D., Megens, H. J. and Li, S. (2012). Analyses of pig genomes provide insight into porcine demography and evolution. *Nature*, 491, 393-8.

Hansson, M., Lundeheim, N., Nyman, G. and Johansson, G. (2011). Effect of local anaesthesia and/or analgesia on pain responses induced by piglet castration. *Acta Veterinaria Scandinavica*, 53(34), 1-9.

Harley, S., Boyle, L. A., O'Connell, N. E., More, S., Teixeira, D. and Hanlon, A. (2014). Docking the value of pig meat? Prevalence and financial implications of welfare lesions in Irish slaughter pigs. *Animal Welfare*, 23, 275-85.

Hartsock, T. G. and Graves, H. B. (1976). Neonatal behavior and nutrition-related mortality in domestic swine. *Journal of Animal Science*, 42(1), 235-41.

Hay, M., Rue, J., Sansac, C., Brunel, G. and Prunier, A. (2004). Long-term detrimental effects of tooth clipping or grinding in piglets: a histological approach. *Animal Welfare*, 13(1), 27-32.

Heim, G., Mellagi, A. P., Bierhals, T., de Souza, L. P., de Fries, H. C., Piuco, P., Seidel, E., Bernardi, M. L., Wentz, I., Bortolozzo, F. P. (2012). Effects of cross-fostering within 24h after birth on pre-weaning behaviour, growth performance and survival rate of biological and adopted piglets. *Livestock Science*, 31, 150(1), 121-7.

Herpin, P., Damon, M. and Le Dividich, J. (2002). Development of thermoregulation and neonatal survival in pigs. *Livestock Production Science*, 78(1), 25-45.

Herskin, M. S., Thodberg, K. and Jensen, H. E. (2015). Effects of tail docking and docking length on neuroanatomical changes in healed tail tips of pigs. *Animal*, 9(04), 677-81.

Hess, A., Boddicker, N., Rowland, R. R. R., Lunney, J. K., Plastow, G. S. and Dekkers, J. C. M. (2014, August). Genetic Parameters and Effects for a Major QTL of Piglets Experimentally Infected with a Second Porcine Reproductive and Respiratory Syndrome Virus. In Proceedings, 10th World Congress of Genetics Applied to Livestock Production.

Hewson, C. J., Dohoo, I. R., Lemke, K. A. and Barkema, H. W. (2007). Canadian veterinarians' use of analgesics in cattle, pigs, and horses in 2004 and 2005. *The Canadian Veterinary Journal*, 48(2), 155.

Hillier, S. G. (1994).Current concepts of the roles of follicle stimulating hormone and luteinizing hormone in folliculogenesis. *Human Reproduction* 9(2), 188-91.

Hunter, E. J., Jones, T. A., Guise, H. J., Penny, R. H. C. and Hoste, S. (2001). The relationship between tail biting in pigs, docking procedure and other management practices. *The Veterinary Journal*, 161(1), 72-9.

Ijaz, A., Abalkhail, A. A. and Khamas, W. A. H. (2000). Effect of intra testicular injection of formalin on seminiferous tubules in Awassi lambs. *Pakistan Veterinary Journal*, 20(3), 129–34.

Jensen, P. and Redbo, I. (1987). Behaviour during nest leaving in free-ranging domestic pigs. *Applied Animal Behaviour Science*, 18(3–4), 355–62.

Johnson, A. K., Julie, M.-T. and McGlone, J. J. (2001). Behavior and Performance of Lactating Sows and Piglets Reared Indoors or Outdoors. *Journal of Animal Science*, 79, 2588–96.

KilBride, A. L., Mendl, M., Statham, P., Held, S., Harris, M., Marchant-Forde, J. N., Booth, H., and Green, L. E. (2014). Risks associated with Preweaning mortality in 855 litters on 39 commercial outdoor pig farms in England. *Preventive Veterinary Medicine*, 117, 189–99.

King, J. W. B. and Gajić, Z. (1969) The repeatability of maternal performance in inbred, outbred and linecross large white sows. *Animal Science*, 11, 47–51.

Knauer, M., Stalder, K. J., Karriker, L., Baas, T. J., Johnson, C., Serenius, T., Layman, L. and McKean, J. D. A descriptive survey of lesions from cull sows harvested at two Midwestern US facilities. *Preventive Veterinary Medicine*, 14, Dec 2007, 82(3), 198–212.

Kuller, W. I., Soede, N. M., van Beers-Schreurs, H. M. G., Langendijk, P., Taverne, M. A. M., Verheijden, J. H. M. and Kemp, B. (2004). Intermittent suckling: Effects on piglet and sow performance before and after weaning. *Journal of Animal Science*, 82(2), 405–13.

Lambooij, W., van Putten. Transport of pigs. Livestock Handling and Transport. (1993). Granding ed. CAB International, Wallingford, UK, pp. 213–31.

Langendijk, P., Dieleman, S. J., Van Den Ham, C. M., Hazeleger, W., Soede, N. M. and Kemp, B. (2007). LH pulsatile release patterns, follicular growth and function during repetitive periods of suckling and non-suckling in sows. *Theriogenology*, 67(5), 1076–86.

Le Dividch, J. and Seve, B. (2000, September). Energy requirement of the young piglet. The weaner pig: Nutrition and management. In Proceedings of a British Society of Animal Science Occasional Meeting, University of Nottingham, UK.

Lewis, E., Boyle, L. A., Brophy, P., O'doherty, J. V. and Lynch, P. B. (2005). The effect of two teeth resection procedures on the welfare of sows in farrowing crates. Part 2. *Applied Animal Behaviour Science*, 90(3), 251–65.

Lewis, E., Boyle, L. A., Lynch, P. B., Brophy, P. and O'doherty, J. V. (2005). The effect of two teeth resection procedures on the welfare of piglets in farrowing crates. Part 1. *Applied Animal Behaviour Science*, 90(3), 233–49.

Lewis, N. Transport of early weaned pigs. (2008). *Behavioral Science*, 110, 128–35.

Livingston, D. H., Mosenthal, A. C. and Deitch, E. A. (1995). Sepsis and multiple organ dysfunction syndrome: a clinical-mechanistic overview. *New horizons* (Baltimore, Md.), 3(2), 257–66.

Lovell, J. S. (2016). Understanding farm animal abuse. *The Routledge International Handbook of Rural Criminology*, 137.

Lou, Z. and Hurnick, J. F. (1994). An ellipsoid farrowing crate: its ergonomical design and effects on pig productivity. *Journal of Animal Science*, 72, 2610–16.

Lumb, S. (2007) Towards a more 'humane' castration for piglets. *Pig Progress*, 23, 24–6

Main, R. G., Dritz, S. S., Tokach, M. D., Goodband, R. D. and Nelssen, J. L. (2004). Increasing weaning age improves pig performance in a multisite production system. *Journal of Animal Science*, 82(5), 1499-507.

Manteuffel, G. and Schön, P. C. (2002). Measuring pig welfare by automatic monitoring of stress calls. *Bornimer Agrartech. Ber*, 29, 110-18.

Marchant-Forde, J. N., Lay, D. C., McMunn, K. A., Cheng, H. W., Pajor, E. A. and Marchant-Forde, R. M. (2009). Postnatal piglet husbandry practices and well-being: the effects of alternative techniques delivered separately. *Journal of Animal Science*, 87(4), 1479-92.

Martínez-Avilés, M., Fernández-Carrión, E., López García-Baones, J. M. and Sánchez-Vizcaíno, J. M. (2015). Early detection of infection in pigs through an online monitoring system. *Transboundary and emerging diseases*, 64(2), 364-73.

Marx, G., Horn, T., Thielebein, J., Knubel, B. and Von Borell, E. (2003). Analysis of pain-related vocalization in young pigs. *Journal of Sound and Vibration*, 266(3), 687-98.

Maselyne, J., Adriaens, I., Huybrechts, T., De Ketelaere, B., Millet, S., Vangeyte, J., van Nuffel., A. and Saeys, W. (2015). 5.5. Assessing the drinking behaviour of individual pigs using RFID registrations. Ilan, H. (ed.), p. 209.

Maselyne, J., Saeys, W. and Van Nuffel, A. (2015). Review: Quantifying animal feeding behaviour with a focus on pigs. *Physiology and Behavior*, 138, 37-51.

Matthews, S. G., Miller, A. L., Clapp, J., Plötz, T. and Kyriazakis, I. (2016). Early detection of health and welfare compromises through automated detection of behavioural changes in pigs. *The Veterinary Journal*, 217, 43-51.

McCracken, B. A., Gaskins, H. R., Ruwe-Kaiser, P. J., Klasing, K. C. and Jewell, D. E. (1995). Diet-dependent and diet-independent metabolic responses underlie growth stasis of pigs at weaning. *Journal of Nutrition*, 125(11), 2838-45.

McGlone, J. J. (1985). Olfactory cues and pig agonistic behavior: evidence for a submissive pheromone. *Physiology and Behavior*, 34(2), 195-8.

McGlone, J. J. (2002). Housing options for farrowing: Considerations for Animal Welfare and Economics. Purdue Extension. PIH 01-01-02, 6p.

McGlone, J. J. and Blecha, F. (1987). An examination of behavioral, immunological and productive traits in four management systems for sows and piglets. *Applied Animal Behaviour Science*, 18, 269-86.

McGlone, J. J. and Anderson, D. L. (2002). Synthetic maternal pheromone stimulates feeding behavior and weight gain in weaned pigs. *Journal of Animal Science*, 80, 3179-83.

McGlone, J. J. and Hellman, J. M. (1988). Local and general anesthetic effects on behavior and performance of two-and seven-week-old castrated and uncastrated piglets. *Journal of Animal Science*, 66(12), 3049-58.

McGlone, J. J. and Morrow, J. L. (1988). Reduction of pig agonistic behavior by androstenone. *Journal of Animal Science*, 66(4), 880-4.

McGlone, J. J., Akins, C. K. and Green, R. D. 1991. Genetic variation of sitting frequency and duration in pigs. *Applied Animal Behavior*, 30, 319-22.

McGlone, J. J., Nicholson, R. I., Hellman, J. M. and Herzog, D. N. (1993). The development of pain in young pigs associated with castration and attempts to prevent castration-induced behavioral changes. *Journal of Animal Science*, 71(6), 1441-6.

McGlone, J. J., Thompson, W. G. and Guay, K. A. (2014). Case study: The pig pheromone androstenone, acting as an interomone, stops dogs from barking. *The Professional Animal Scientist*, 30(1), 105-8.

McGlone, J. J., Thompson, G., Devaraj, S. G. (2016). A natural interomone 2-methyl-2 butenal stimulates feed intake and weight gain in weaned pigs. *Animal*, 11 (2), 306–8.

McGlone, J. J. (2006). Comparison of sow welfare in the Swedish deep-bedded system and the US crated-sow system. *Journal of American Veterinary Medical Association*, 229, 1377–80.

McKean, J. D. (2001). The importance of traceability for public health and consumer protection. *Revue scientifique et technique* (International Office of Epizootics), 20(2), 363–71.

McManus, C., Tanure, C. B., Peripolli, V., Seixas, L., Fischer, V., Gabbi, A. M., Menegassi, S. R., Stumpf, M. T., Kolling, G. J., Dias, E. and Costa, J. B. G. (2016). Infrared thermography in animal production: An overview. *Computers and Electronics in Agriculture*, 123, 10–16.

Meijer., E., Bertholle, C. P., Oosterlinck, M., vander Staay, F., Back, W. and van Nes, A. (2014). Pressure mat analysis of the longitudinal development of pig locomotion in growing pigs after weaning pigs. *BMC Veterinary Research*, 10, 1–11.

Meiszberg, A. M., Johnson, A. K., Sadler, L. J., Carroll, J. A., Dailey, J. W. and Krebs, N. (2009). Drinking behavior in nursery pigs: Determining the accuracy between an automatic water meter versus human observers. *Journal of Animal Science*, 87(12), 4173–80.

Melin, L., Mattsson, S., Katouli, M. and Wallgren, P. (2004). Development of Post-weaning Diarrhoea in Piglets. Relation to Presence of Escherichia coli Strains and Rotavirus. *Journal of Veterinary Medicine*, Series B, 51(1), 12–22.

Merks, J. W. M. and Lambooy, E. (1990). Injectable electronic identification systems in pig production. *Pig News and Information*, 11(1), 35–6.

Moeser, A. J., Vander Klok, C., Ryan, K. A., Wooten, J. G., Little, D., Cook, V. L. and Blikslager, A. T. (2007). Stress signaling pathways activated by weaning mediate intestinal dysfunction in the pig. *American Journal of Physiology-Gastrointestinal and Liver Physiology*, 292(1), G173–81.

Morrow-Tesch, J. and McGlone, J. J. (1990). Sources of maternal odors and the development of odor preferences in baby pigs. *Journal of Animal Science*, 68, 3563–71.

Mucignat-Caretta, C. (ed.), (2014). *Neurobiology of Chemical Communication*. CRC Press.

Nguyen, D. T., Lee, K., Choi, H., Choi, M., Le, M. T., Song, N., Kim, J., Seo, H. G., Oh, J. W., Lee, K., Kim, T. and Park, C. (2012). The complete swine olfactory subgenome: expansion of the olfactory gene repertoire in the pig genome. *BMC Genomics*, 13, 584–96.

Niekamp, S. R., Sutherland, M. A., Dahl, G. E. and Salak-Johnson, J. L. (2007). Immune responses of piglets to weaning stress: Impacts of photoperiod. *Journal of Animal Science*, 85(1), 93–100.

Noonan, G. J., Rand, J. S., Priest, J., Ainscow, J. and Blackshaw, J. K. (1994). Behavioural observations of piglets undergoing tail docking, teeth clipping and ear notching. *Applied Animal Behaviour Science*, 39(3–4), 203–13.

Norman, A. W. and Litwack, G. (1997). *Hormones*, New York: Academic Press. New York.

Nuntapaitoon, M. and P. Tummaruk. (2015). Piglet Preweaning mortality in a commercial swine herd in Thailand. *Tropical Animal Health and Production*. 47, 1539–46.

Pageat, P. (2001). Pig appeasing pheromone to decrease stress, anxiety and aggressiveness. USA Patent 6,169,113. Publication date 2 January 2001.

Paoli, M. A., Lahrmann, H. P., Jensen, T. and D'Eath, R. B. (2016). Behavioural differences between weaner pigs with intact and docked tails. *Animal Welfare*, 25(2), 287–96.

Pederson, L. J., Jensen, H. and Thodberg, K. (2008). Cross fostering of piglets in farrowing pens. Proceedings of the 20th Nordic Symposium of the International Society of the International Society of Applied Ethology. Norway.

Perez-Pinera, P., Ousterout, D. G. and Gersbach, C. A. Advances in Targeted Genome Editing. (2012). *Current Opinion in Chemical Biology*, 16, 268–77.

Plastow, G. S. (2016). Genomics to benefit livestock production: improving animal health. *Revista Brasileira de Zootecnia*, 45(6), 349–54.

Price, E. O., Hutson, G. D., Price, M. I. and Borgwardt, R. (1994). Fostering in swine as affected by age of offspring. *Journal of Animal Science*, 72(7), 1697–701.

Prunier, A., Bonneau, M., Von Borell, E. H., Cinotti, S., Gunn, M., Fredriksen, B., Giersing, M., Morton, D. B., Tuyttens, F. A. and Velarde, A. (2006). A review of the welfare consequences of surgical castration in piglets and the evaluation of non-surgical methods. *Animal Welfare-Potters Bar then Thampstead*. August 1, 15(3), 277.

Prunier, A., Mounier, A. M. and Hay, M. (2005). Effects of castration, tooth resection, or tail docking on plasma metabolites and stress hormones in young pigs. *Journal of Animal Science*, 83(1), 216–22.

Prunier, A., Mounier, A. M., Bregeon, A. and Hay, M. (2001). Influence of tail docking, tooth resection and castration on plasma cortisol, ACTH, glucose and lactate in piglets. In Proceedings of the 4th International Conference on Farm Animal Endocrinology, Parme, Italy, 7–10 October 2001.

Rault, J. L. and Lay, D. C. (2011). Nitrous oxide by itself is insufficient to relieve pain due to castration in piglets. *Journal of Animal Science*, 89, 3318–25.

Rault, J. L., Lay, D. C. and Marchant-Forde, J. N. (2011). Castration induced pain in pigs and other livestock. *Animal Behaviour Science*, 135(3), 214–25.

Rea, C. P. (1986). Universal ear notching system in swine. Agricultural Guide. Swine Management. University of Missouri-Colombia Extension. 2505.

Rekwot, P. I., Ogwu, D., Oyedipe, E. O. and Sekoni, V. O. (2001). The role of pheromones and biostimulation in animal reproduction. *Animal Reproduction Science*, 65(3), 157–70.

Robert, S. and Martineau, G. P. (2001). Effects of repeated cross-fosterings on preweaning behavior and growth performance of piglets and on maternal behavior of sows. *Journal of Animal Science*, 79(1), 88–93.

Robert, S., Thompson, B. K. and Fraser, D. (1995). Selective tooth clipping in the management of low-birth-weight piglets. *Canadian Journal of Animal Science*, 75(3), 285–9.

Robertson, J. B., Laird, R., Hall, J. K. S., Forsyth, R. J., Thomson, J. M. and Walker-Love, J. (1966) 'A comparison of two indoor farrowing systems for sows', *Animal Science*, 8, 171–8.

Rushen, J., Chapinal, N. and De Passille, A. M. (2012). Automated monitoring of behavioural-based animal welfare indicators. *Animal Welfare-The UFAW Journal*, 21(3), 339.

Rusk, C. P., Blomeke, C. R., Balschweid, M. A., Elliot, S. J. and Baker, D. (2006). An evaluation of retinal imaging technology for 4-H beef and sheep identification. *Journal of Extension*, 44(5), 1–33.

Sandercock, D. A., Smith, S. H., Di Giminiani, P. and Edwards, S. A. (2016). Histopathological Characterization of Tail Injury and Traumatic Neuroma Development after Tail Docking in Piglets. *Journal of Comparative Pathology*, 155, 276.

Scaramuzzi, R. J. and Martin, G. B. (2008). The importance of interactions among nutrition, seasonality and socio-sexual factors in the development of hormone-free methods for controlling fertility. *Reproduction in Domestic Animals*, 43, 129-36.

Schneider, F., Falkenberg, H., Kuhn, G., Nürnberg, K., Rehfeldt, C. and Kanitz, W. (1998). Effects of treating young boars with a GnRH depot formulation on endocrine functions, testis size, boar taint, carcass composition and muscular structure. *Animal Reproduction Science*, 50(1), 69-80.

Schön, P. C., Puppe, B. and Manteuffel, G. (2004). Automated recording of stress vocalisations as a tool to document impaired welfare in pigs. *Animal Welfare*, 13(2), 105-10.

Scott, K., Taylor, L., Gill, B. P. and Edwards, S. A. (2006). Influence of different types of environmental enrichment on the behaviour of finishing pigs in two different housing systems: 1. Hanging toy versus rootable substrate. *Applied Animal Behaviour Science*, 99(3), 222-9.

Sørensen, J. T., Rousing, T., Kudahl, A. B., Hansted, H. J. and Pedersen, L. J. (2016). Do nurse sows and foster litters have impaired animal welfare? Results from a cross-sectional study in sow herds. *Animal*, 10(04), 681-6.

Spencer, J. D., Boyd, R. D., Cabrera, R. and Allee, G. L. (2003). Early weaning to reduce tissue mobilization in lactating sows and milk supplementation to enhance pig weaning weight during extreme heat stress. *Journal of Animal Science*, 81(8), 2041-52.

Spreeuwenberg, M. A. M., Verdonk, J. M. A. J., Gaskins, H. R. and Verstegen, M. W. A. (2001). Small intestine epithelial barrier function is compromised in pigs with low feed intake at weaning. *The Journal of Nutrition*, 131(5), 1520-7.

Støier, S., Larsen, H. D., Aaslyng, M. D. and Lykke, L. (2016). Improved animal welfare, the right technology and increased business. *Meat Science*.

Sutherland, M. A. and Tucker, C. B. (2011). The long and short of it: A review of tail docking in farm animals. *Applied Animal Behaviour Science*, 135(3), 179-91.

Sutherland, M. A., Bryer, P. J., Krebs, N. and McGlone, J. J. (2008). Tail docking in pigs: acute physiological and behavioural responses.

Sutherland, M. A., Bryer, P. J., Krebs, N. and McGlone, J. J. (2009). The effect of method of tail docking on tail-biting behaviour and welfare of pigs. *Animal Welfare*, 18(4), 561-70.

Sutherland, M. A., Davis, B. L. and McGlone, J. J. (2011). The effect of local or general anesthesia on the physiology and behavior of tail docked pigs. *Animal*, 5(08), 1237-46.

Sutherland, M. A., Backus, B. L. and McGlone, J. J. (2014) Effects of transport at weaning on the behavior, physiology and performance of pigs. *Animal*, 4, 657-69.

Tao, X., Xu, Z. and Men, X. (2016). Transient effects of weaning on the health of newly weaning piglets. *Czech Journal of Animal Science*, 61(2), 82-90.

Thodberg, K., Jensen, K. H. and Jørgensen, E. (2010). The risk of tail-biting in relation to level of tail-docking. In Proceedings of the 44th Congress of the International Society for Applied Ethology (isae).

Thompson, B. K. and Fraser, D. (1986). Variation in piglet weights: development of within-litter variation over a 5-week lactation and effect of farrowing crate design. *Canadian Journal of Animal Science*, 66(2), 361–72.

Tirindelli, R., dibattista, M., Pifferi, S. and Menini, A. From pheromones to behavior. 2009. *Physiological Reviews*, 89, 921–56.

Tokach, M. D., Goodband, R. D., Nelssen, J. L., Kats, L. J., Goodband, B. and Tokach, M. (1992). Influence of weaning weight and growth during the first week postweaning on subsequent pig performance. In Kansas State University Swine Day 1992. Report of Progress 667. Kansas State University, pp. 19–21.

Turpin, D. L., Langendijk, P., Chen, T. Y., Lines, D. and Pluske, J. R. (2016). Intermittent Suckling Causes a Transient Increase in Cortisol That Does Not Appear to Compromise Selected Measures of Piglet Welfare and Stress. *Animal*, 6(3), 24.

Van de Weerd, H. A., Docking, C. M., Day, J. E., Avery, P. J. and Edwards, S. A. (2003). A systematic approach towards developing environmental enrichment for pigs. *Applied Animal Behaviour Science*, 84(2), 101–18.

van Staaveren, N., Teixeira, D. L., Hanlon, A. and Boyle, L. A. (2016). Pig carcass tail lesions: the influence of record keeping through an advisory service and the relationship with farm performance parameters. *Animal*, 1–7.

Viazzi, S., Ismayilova, G., Oczak, M., Sonoda, L. T., Fels, M., Guarino, M., Vranken, E., Hartung, J., Bahr, C. and Berckmans, D. (2014). Image feature extraction for classification of aggressive interactions among pigs. *Computers and Electronics in Agriculture*, 104, 57–62.

Von Borell, E., Baumgartner, J., Giersing, M., Jäggin, N., Prunier, A., Tuyttens, F. A. M. and Edwards, S. A. (2009). Animal welfare implications of surgical castration and its alternatives in pigs. *Animal*, 3(11), 1488–96.

Voulodimos, A. S., Patrikakis, C. Z., Sideridis, A. B., Ntafis, V. A. and Xylouri, E. M. (2010). A complete farm management system based on animal identification using RFID technology. *Computers and Electronics in Agriculture*, 70(2), 380–8.

Wayne, I. P. F., Wayne, F., Rusk, I. D. C. P. and Richert, B. T. Identification of Swine by Auricular Vein Patterns. https://pdfs.semanticscholar.org/5c2e/df3cadb5eb0cd139167992976eb1b6d4e1e7.pdf. Accessed April 17, 2017.

Wemelsfelder, F. and Putten, V. G. (1985). Behaviour as a possible indicator for pain in piglets. I.V.O.-report. Instituut voor Veeteeltkundig Onderzoek 'Schoonoord' no. B-260.

Westin, R., Holmgren, N., Hultgren, J., Ortman, K., Linder, A. and Algers, B. (2015). Post-mortem findings and piglet mortality in relation to strategic use of straw at farrowing. *Preventive Veterinary Medicine*, 119, 141–52.

White, R. G., DeShazer, J. A., Tressler, C. J., Borcher, G. M., Davey, S., Waninge, A., Parkhurst, A. M., Milanuk, M. J. and Clemens, E. T. (1995). Vocalization and physiological response of pigs during castration with or without a local anesthetic. *Journal of Animal Science*, 73, 381–6.

Widowski, T. and Torrey, S. (2002). Swine Fact Sheet. http://csm.pork.org/filelibrary/Factsheets/Well-Being/SWINE%20WELFAREFACTSHT-neona.pdf. Accessed April 17, 2017.

Zijlstra, R. T., Whang, K. Y., Easter, R. A. and Odle, J. (1996). Effect of feeding a milk replacer to early-weaned pigs on growth, body composition, and small intestinal morphology, compared with suckled littermates. *Journal of Animal Science*, 74(12), 2948-59.

Zonderland, J. J., Wolthuis-Fillerup, M., Van Reenen, C. G., Bracke, M. B., Kemp, B., Den Hartog, L. A. and Spoolder, H. A. (2008). Prevention and treatment of tail biting in weaned piglets. *Applied Animal Behaviour Science*, 110(3), 269-81.

Zurbrigg, K. (2006). Sow shoulder lesions: Risk factors and treatment effects on an Ontario farm. *Journal of Animal Science*, 84(9), 2509-14.

Chapter 4

Optimising the health of weaned piglets

Andrea Luppi, Istituto Zooprofilattico Sperimentale della Lombardia e dell'Emilia Romagna (IZSLER), Italy

1 Introduction

The immediate post-weaning period is one of the most stressful phases in a pig's life in which piglets are usually exposed to environmental, social and psychological stressors such as abrupt separation from their mother, mixing with other litters in a usually new environment and switching from highly digestible (liquid) milk to a less-digestible plant-based dry diet containing complex protein and carbohydrate, including various antinutritional factors (Lallès et al., 2007). All these elements have direct or indirect effects on health and overall growth performance.

At weaning several factors influence the health and performance of pigs (Fig. 1), determining a suboptimal weight gain, predisposing weaned pigs to the disease, such as episodes of diarrhoea, and increased morbidity and mortality (Pluske et al., 1997) and reduced growth performance (Vente-Spreeuwenberg et al., 2003).

The immediate post-weaning period is also characterised by the immaturity of the digestive system, predisposing the piglet to diseases (Jayaraman and

http://dx.doi.org/10.19103/AS.2022.0103.17

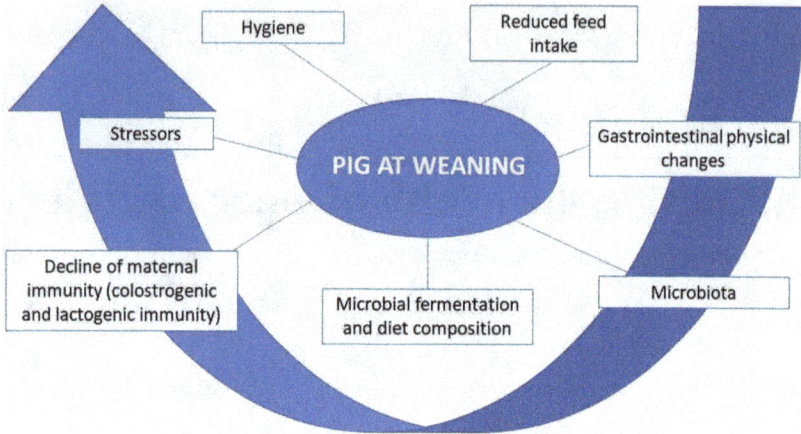

Figure 1 Main factors influencing health and performance of pigs at weaning (Pluske et al., 1997; Jayaraman and Nyachoti, 2017).

Nyachoti, 2017). In this phase, the immaturity of the intestinal immune system (complicated by the lack of a previous stimulation by pathogens) may reduce the ability of pigs to mount an appropriate immunological response to pathogens (Heo et al., 2013).

Piglets are born agammaglobulinemic and do not become fully immunologically competent until about 4 weeks of age (Langel et al., 2016). Maternal immunity, composed of colostrogenic and lactogenic immunity, is able to modulate the immune response and provide specific immunity in the newborn (Poonsuk and Zimmerman, 2018). More specifically the maternal immune components are characterised by: (1) circulating antibodies derived from colostrum; (2) mucosal antibodies from colostrum and milk; and (3) immune cells provided in mammary secretions (Poonsuk and Zimmerman, 2018). At weaning, in general terms, the protection given by the colostrogenic and lactogenic immunity ends. The severity of this condition is dependent on how much the immune system has developed during the pre-weaning period (Bailey et al., 2005) exposing piglets to several infections and possible diseases.

Weaning weight is an important factor able to influence the pig's growth and feed efficiency from weaning to market. Colostrum intake affects weaning weight showing long-term effects on piglets' growth from 3 weeks of age until after weaning (Devillers et al., 2011). Extending weaning age is a very effective way of improving weaning weight (Pinilla et al., 2008). This is very important considering that the weaning age imparted by a swine production system does not match with the natural situation, where weaning is a gradual process that takes place between 2 and 3 months of age (Allwin et al., 2016).

Another important factor associated with high variability in pig weight at weaning is the within-litter variation of piglet birth weight (Quesnel et al., 2008).

An increase in litter size in hyperprofilic sows was shown to be genetically associated with a decrease in the mean piglet birth weight and an increase in the within-litter variability of birth weight (Wolf et al., 2008).

Improving husbandry practices in the swine barn, in order to reduce the negative impact of the weaning transition period, include nutrition strategies, management practices, maintenance of hygienic standards, disease prevention protocols and animal welfare considerations (e.g. formation of groups, social stressors and ambient temperature for piglets) (Jayaraman and Nyachoti, 2017).

2 Transition from weaning: factors influencing health and performance of pigs

2.1 Reduced feed intake

The period of transition from weaning and the first post-weaning period are characterised by a marked reduction in voluntary feed intake. Lallès et al. (2007) reported that although 50% of weaned piglets consume their first meal within 24 h post-weaning, 10% have not eaten until 48 h. Thus, energy requirements for maintenance are only met 3 days post-weaning, and it can take 8–14 days for piglets to recover their pre-weaning level of energy intake. Available evidences indicate that reduced feed intake is a major contributing factor to the abruptly reduced intestinal villus height, affecting gut morphology, barrier function and the overall efficiency of nutrient capture and utilisation (Pluske et al., 1997; Vente-Spreeuwenberg et al., 2001; Vente-Spreeuwenberg et al., 2003). Feed intake and its reduction can predispose pigs to enteric diseases and is influenced by various factors (Jayaraman and Nyachoti, 2017) (Fig. 2).

In swine husbandry, practising encouraging feed intake and reducing weaning stresses are important factors for maintaining performance, the integrity of the small intestinal structure immediately after weaning (Moeser et al., 2017) and reducing mortality and morbidity (Collins et al., 2017). During the post-weaning phase psychological stressors, such as mixing with other littermates (Khafipour et al., 2014), crowding stress, and environmental stressors, such as hygienic standards, air quality and change in ambient temperature, act to drastically reduce feed intake (Campbell et al., 2013). Providing adequate feeder space is critical to weaned pig performance because limited feeder space can increase competition at the feeder, compromising feed intake and leading to reduced growth rates (Averos et al., 2012). He et al. (2016) evaluated the effect of restricted feeder space (2 spaces per pen vs. 5 spaces per pen; the area of each feeding space was 15 cm × 15 cm) on the growth performance of weaned pigs. The authors indicated that limited feeder space is associated with an increased risk of mortality or slow growth in weaned pigs.

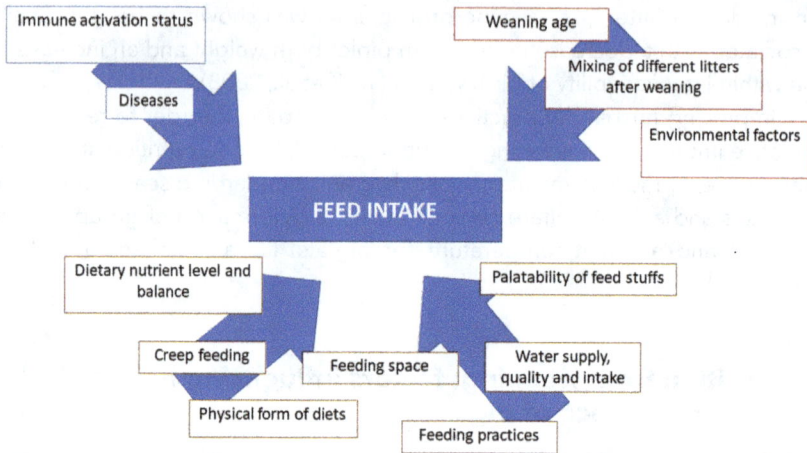

Figure 2 Main factors affecting feed intake at weaning (Jayaraman and Nyachoti, 2017).

Water intake might influence feed intake. Dybkjær et al. (2006) reported that a low feed intake may be due to insufficient drinking activity, as solid feed intake must be accompanied by water intake.

There is generally a negative relationship between inadequate thermal comfort (air temperature, relative humidity and ventilation rate) and feed intake (Noblet et al., 2001). Recommended air temperature ranges at the animal level and minimum ventilation rates for pigs of various sizes and ages are reported in Tables 1 and 2.

Table 1 Recommended air temperature ranges for pigs of various sizes and ages (Zulovich, 2012)

Pig's category	Optimal temperature (°C)	Range (°C)
Litter-newborn	35	32–35
Litter-3 weeks old	27	24–29
Nursery 5-14 kg	27	24–29
Nursery 14-23 kg	24	21–27
Nursery 23-34 kg	18	16–21

Table 2 Target minimum ventilation rates for pigs of various sizes (modified from Zulovich, 2012)

Pig's category	Ventilation rates
Litter	0.564 m³/per minute-crate (20 cfm/crate)
Nursery 4-14 kg	0.0282–0.0564 m³/per minute-pig (1-2 cfm/pig)
Nursery 14-23 kg	0.084–0.114 m³/per minute-pig (3-4 cfm/pig)
Nursery 23-34 kg	0.144–0.198 m³/per minute-pig (5-7 cfm/pig)

cfm, cubic feet per minute.

About the change in ambient temperature, fluctuations of ±4°C during the first week after weaning have been demonstrated to increase post-weaning scours (Brumm, 2019).

2.2 Gastrointestinal physical changes

Weaned pigs have a higher gastric pH value and this may be partly due to a lower acid secretion capacity of the stomach at weaning along with the reduction in lactic acid production from lactose. The high gastric pH value after weaning may contribute, in part, to the susceptibility of piglets to enteric infections at this time (Heo et al., 2013), considering that exposure to low pH values (i.e. 3.0–4.0) is bactericidal for many pathogenic bacteria, including *Escherichia coli*.

Contributing factors that increase gastric pH at weaning are:

- lower (hydrochloric acid) HCl production,
- abrupt change in diet,
- overeating after anorexia,
- protein level,
- dietary electrolytes balance,
- dietary concentration of lactose or lactogenic carbohydrates,
- lower secretion of saliva.

Weaning also reduces gastric motility, with a reduction in stomach emptying rate in pigs after weaning compared with suckling pigs. Martinez et al. (2004) reported that the activation of stress gene, corticotrophin-releasing factor receptor 2, has been implicated in the inhibition of gastric motility, while Moeser et al. (2007) described its up-regulation in the jejunum of weaned pigs. Gastric stasis may contribute to the development of post-weaning diarrhoea (PWD) in piglets by allowing the proliferation of pathogenic bacteria (Heo et al., 2013).

During the post-weaning period, changes in intestinal tissue, including villus and crypt architecture can be associated with depressed activities of many brush-border digestive enzymes and reduced absorptive capacity (Hampson and Kidder, 1986; Nabuurs et al., 1993; Vente-Spreeuwenberg et al., 2004). For this reason, activities of the brush-border enzymes in weaned pigs have been used as indicators of maturation and digestive capacity of the small intestine (Heo et al., 2013). The major effects of an immature digestive system in weaned pigs include the reduced activity of digestive enzymes, changes in intestinal morphology (Boudry et al., 2004) and reduced digestion in the small intestine.

2.3 Microbiota

An important topic involving piglets in the first weeks of life and in the weaning transition period is about the intestinal microbiota. The gut is sterile at birth and is then colonised by bacteria from the mother and the environment, starting with lactic acid bacteria, enterobacteria and streptococci. After the introduction of solid feed, obligate anaerobes increase in number and diversity until an adult-type pattern is achieved (Lallès et al., 2007). In a short period of time, the intestinal microbiota must ultimately develop from a simple unstable community into a complex and stable community, thus generating a tight 'colonisation resistance' or 'competitive exclusion' (Lallès et al., 2007). Microbial changes have been reported in the first 7-14 days after weaning when the number of lactobacilli diminished regardless of the weaning age (Franklin et al., 2002). It was also observed that lactobacilli numbers decreased whereas coliform and *E. coli* numbers increased in the various regions of the gastrointestinal tract at 28 days after weaning (Konstantinov et al., 2006).

One of the most important factors that affect the total population and diversity of the intestinal microbiota is the diet composition (Castillo et al., 2007; Metzler et al., 2009). Jeaurond et al. (2008) reported that the counts of *Clostridium* spp. in the large intestine were low in pigs fed higher fermentable carbohydrates (such as sugar beet pulp) and tended to increase by increasing dietary levels of fermentable protein.

2.4 Microbial fermentation and diet composition

It is recognised that microbial fermentation within the gastrointestinal tract is very important for the pig (Williams et al., 2001). Fermentation of carbohydrates, specifically resistant starches and dietary fibre, leads to the production of mainly short-chain volatile fatty acids (SCFAs) (acetic, propionic and butyric acids) and the use of NH_3 and other nitrogenous compounds, which are required for microbial growth (Lallès et al., 2007). When carbohydrates are in short supply relative to the available protein of non-degradable and endogenous origin, protein will be used as an energy source for fermentation, resulting in the end products NH_3, branched-chain fatty acids (BCFAs) and potentially toxic end products (Williams et al., 2001). These compounds include amines, volatile phenols and indoles, contributing to the increase in the incidence of diarrhoea at weaning in pigs (Lallès et al., 2007; Riaz Rajoka et al., 2017).

In recent years, the recommended levels of amino acids have increased and the National Research Council (NRC) reports crude protein (CP) concentrations of 20-23% (200-230 g/kg) in the diet of weaned pigs (NRC, 2012). Veldkamp and Vernooij (2021) reported that to maximise piglet growth, it is recommended that levels of CP be at 20-23% in pre-starter offered to piglets, before and just after weaning, and 18–20% in starter diets. There is evidence that feeding with such a high protein diet immediately after weaning could cause protein maldigestion as the digestive system of weaner pigs is insufficiently developed to fully digest and absorb dietary proteins. Consequently, increasing amounts of undigested CP materials are present in the large intestine where it is fermented by resident micro-organisms and could encourage the growth of nitrogen utilising bacteria (Heo et al., 2013; O'Doherty et al., 2017). This contributes to increased production of toxic by-products such as BCFAs, indole, phenols, ammonia and biogenic amines in the gastrointestinal tract (Pluske et al., 2002). Reducing dietary CP is a strategy to reduce the incidence of diarrhoea (Wu et al., 2015). A low-protein diet (<180 g/kg) in the post-weaning period reduces protein fermentation in the gastrointestinal tract (Nyachoti et al., 2006). Opapeju et al. (2009; 2010) showed as diets containing 170 g/kg CP can reduce inflammatory responses and the number of enterotoxigenic *E. coli* (ETEC) in the small and large intestine. Lordelo et al. (2008) reported that a low-protein diet (170 g/kg CP) did not compromise the growth of piglets if they receive similar digestible energy contents and adequate levels of essential amino acids (i.e. lysine, methionine, tryptophan and threonine) including valine and isoleucine. Rattigan et al. (2020) in a recent study showed as dietary CP can be reduced to 18% without compromising growth performance if the diet is correctly supplemented with the essential amino acids within the recommended range for maximising growth performance of weaned pigs. The reduction of CP in the diet with the aforementioned modalities and levels is a strategy currently adopted in many farms.

Dietary fibre has been reported to improve gut health and decrease diarrhoea incidence in pigs (Molist et al., 2009). In particular feeding weaned pigs high-insoluble-fibre diets might better protect them against pathogenic bacteria by increasing the villus length (McRorie and McKeown, 2017). The impact of dietary fibre on piglets' nutrition might be determined by the properties of fibre and/or fibre sources (Lindberg, 2014). For example, the fibre in the wheat bran diet was adapted by piglets and acted as prebiotics (Chen et al., 2013) increasing feed intake and the development of the gastrointestinal tract. A wheat bran diet has also been shown to decrease the amount of pathogenic *E. coli* in the faeces and reduce the incidence of PWD (Molist et al., 2009).

The electrolyte imbalance in post-weaning diets exhibited alterations in motility, changes in paracellular permeability, loss of absorption surface, a change in electrolyte fluxes in post-weaning piglets and, finally, induced PWD (Saravanan et al., 2013). Guzman-Pino et al. (2015) reported that the balance between cation (Na+) and anions (K+/Cl–) and the acid or alkaline load from the diet may strongly alter the acid-base status and growth performance of weaned piglets, while Gao et al. (2019) described as an excess of chloride

Table 3 Minimum mineral levels in nursery diets (NRC, 2012)

	Nursery pigs – weight	
	5–7 kg/BW	7–11 kg/BW
Minerals	Requirements (% or mg/kg of diet)	
Calcium (%)	0.85	0.80
STTD phosphorus (%)	0.45	0.40
ATTD phosphorus (%)	0.41	0.36
Total phosphorus (%)	0.70	0.65
Sodium (%)	0.40	0.35
Chloride (%)	0.50	0.45
Magnesium (%)	0.04	0.04
Potassium (%)	0.30	0.28
Copper (mg/kg)	6	6
Iodine (mg/kg)	0.14	0.14
Iron (mg/kg)	100	100
Manganese (mg/kg)	4	4
Selenium (mg/kg)	0.30	0.30
Zinc (mg/kg)	100	100

STTD, standardised total tract digestible; ATTD, apparent total tract digestibility. The ATTD of phosphorus is calculated based on the phosphorus content in the diet and feaces. Because the phosphorus in the faeces includes phosphorus originating from the body as well as phosphorus derived from the diet, the ATTD of phosphorus is less than the real phosphorus digestibility. The major issue in the use of ATTD of phosphorus is that the values may not be always additive in mixed diets (NRC, 2012). The STTD of phosphorus takes into account the basal endogenous loss of phosphorus originating from the body (Almeida and Stein, 2010).

ions induces a negative dietary electrolyte balance and reduces the growth performance of weaned pigs.

Maintaining an appropriate dietary electrolyte balance (dEB) is critical for pigs to achieve optimal growth performance. Traditionally, the optimal dEB for pigs is reported to be approximately 250 mEq/kg (NRC, 2012).

Jones et al. (2019) reported that increasing dietary dEB up to 243 mEq/kg in nursery diets (from 0 to 8 days after weaning) improved the growth performance of weanling pigs. Minerals requirement estimates for nursery pigs (NRC, 2012) are reported in Table 3.

2.5 Decline of maternal immunity: colostrogenic and lactogenic immunity

Newborn piglet is immunodeficient at birth and, after exposure to infectious agents, takes 7-10 days for a primary antibody or cell-mediated immune response to develop. During this time, resistance to infection depends on the actions of the innate defence mechanisms and the antibody or immune components passively transferred from the sow to the piglet through colostrum and milk (Chase and Lunney, 2019). More specifically the maternal immune components are characterised by (1) circulating antibodies derived from colostrum; (2) mucosal antibodies from colostrum and milk; and (3) immune cells provided in mammary secretions (Poonsuk and Zimmerman, 2018). From 3 days of age until the end of lactation, IgA is the predominant Ig found in sow milk. The majority of milk Ig is synthesised in the mammary gland, whereas colostral Ig is mostly derived from serum, although this varies by Ig class (Chase and Lunney, 2019). At weaning the protection given by the colostrogenic and lactogenic immunity ends. The severity of this condition is dependent on how much the immune system was developed during the pre-weaning period (Bailey et al., 2005) also considering that the weaning age imparted by a swine production system does not match with the immune system capacity for optimal disease prevention (Chase and Lunney, 2019). The development of immuno-competence, namely the ability to mount appropriate responses to antigens, including the ability to generate tolerance to food and commensal bacterial antigens as well as active immune responses to pathogens, is an absolute requirement for optimum growth and performance (Bailey et al., 2001).

Maternal antibody half-life in pigs ranges from 11.3 days to 20 days. Specific examples for half-life for swine maternal antibodies include 16.2 days for Porcine Reproductive and Respiratory Syndrome virus (PRRSV) (Yoon et al., 1995), 14 days for Swine Influenza A viruses (swIAV) (Fleck and Behrens, 2002), 19 days for Porcine Circovirus Type 2 (PCV2) (Opriessnig et al., 2004), 11.3 days for Pseudorabies virus (PRV) (Müller et al., 2005), 11 days for classical swine fever virus (CSFV) (Müller et al., 2005), 20 days for Parvovirus (PPV) (Paul et al., 1982) and 15.8 days for *Mycoplasma hyopneumoniae* (Morris et al., 1994).

Measured by a serological test, maternal antibody levels against different etiological agents fall below the test cut-off level at different times in the pig's life (Table 4).

To measure and know the level of maternal immunity for different pathogens is of fundamental importance to set the vaccination interventions. The timing of vaccination has to consider when the level of maternal antibodies is low enough for an active immune response to progress sufficiently to provide vaccine immunity. Antibody levels often decay to a level still high enough to block responses to vaccine but not high enough to resist a field infection, which creates a window of opportunity for infecting organisms. For these reasons one of the major challenges in developing an active immune response in young pigs has been the interference from maternal immunity (Rose and Andraud, 2017).

Table 4 Maternal immunity duration of different etiological agents in pig diseases

Disease/etiological agent	Maternal antibody persistence (weeks)	Reference
Aujeszky's disease (SHV-1)	14-15	Thomas et al. (2019)
Porcine respiratory and reproductive syndrome (PRRSV)	6	Silva et al. (2015)
Swine influenza (swIAV)	4-14	Van Reeth and Vincent (2019)
Circovirosis (PCV2)	4.9-11.1	Opriessnig et al. (2004)
Parvovirosis (PPV)	20	Truyen and Streck (2019)
Rotavirosis (RVs)	2	Shepherd et al. (2019)
Classical swine fever (CSFV)	8-12	Kirkland et al. (2019)
Encephalomyocarditis (EMCV)	8	Alexandersen et al. (2019)
Porcine cytomegalovirus (PCMV)	8	Thomas et al. (2019)
Foot and mouth disease (FMDV)	8-12	Alexandersen et al. (2019)
Vomiting and wasting disease (pHEV)	4-18	Saif et al. (2019)
PED (PEDV)	4	Saif et al. (2019)
Hepatitis E (HEV)	7-9	Meng et al. (2019)
Porcine lymphotropic herpesviruses (PLHVs)	3	Thomas et al. (2019)
Mycoplasmosis (*Mycoplasma hyopneumoniae*)	4-8	Maes et al. (1996)
Pleuropneumonia (*Actinobacillus pleuropneumoniae*)	5-12	Gottschalk and Broes (2019)
Glässer's disease (*Glaesserella parasuis*)	3-6	Aragon et al. (2019)
Erysipelas (*Erysipelthrix rushiopathiae*)	8	Opriessnig and Coutinho (2019)
Streptococcosis (*Streptococcus suis*)	6	Gottschalk and Segura (2019)
Bordetellosis (*Bordetella bronchiseptica*)	3-4	Brockmeier et al. (2019)
Enteric colibacillosis (Enterotoxigenic *Escherichia coli*)	3-4	Melkebeek et al. (2013)
Oedema disease (Shiga toxin-producing *E. coli*)	3-4	Melkebeek et al. (2013)
Salmonellosis (*Salmonella* Choleraesuis and *Salmonella* monophasic and biphasic Typhimurium)	Up to 10	Cevallos-Almeida et al. (2019)
Ileitis (*Lawsonia intracellularis*)	3-5	Karuppannan and Opriessnig (2018)

2.6 Stressors

At weaning, the piglet is exposed to various stressors (such as removal of sow's milk, mixing with other littermates and change in ambient temperature) and pathogenic agents, mingling with unfamiliar animals, unfavourable environmental changes, contaminated air and low biosecurity measures. The combined and additive effects of these stressors and pathogenic threats accelerate the detrimental effects on the growth performance of pigs (Hyun et al., 1998).

Khafipour et al. (2014) reported that crowding stress had a direct impact on immune response in pigs as indicated by an elevated serum cytokine and plasma cortisol. Crowding stress leads also to an increase in intestinal pH, which inhibits beneficial bacteria and may create favourable conditions for ETEC colonisation and other harmful bacteria, resulting in diarrhoea (Ohland and MacNaughton, 2010).

2.7 Hygiene

Under poor sanitary conditions, piglets had depressed growth performance. The stimulation of the immune system and the consequent inflammatory response interferes with growth because of competition for nutrients, in particular amino acids, between structural tissues and immune function (Le Floc'h et al., 2009; Jayaraman et al., 2015). About amino acids, the optimal standardised ileal digestible threonine:lysine for weaned pigs, raised under unclean sanitary conditions, should be higher compared with weaned pigs raised under clean sanitary conditions (Jayaraman et al., 2015). Maintaining high standards of cleanliness in the nursery is critical for the optimal performance of piglets partly because of its direct or indirect effects on gut health and function. It was reported that piglets raised under unclean conditions had shorter villous height and less crypt depth compared to those raised under clean conditions (Jayaraman and Nyachoti, 2017).

Jayaraman et al. (2017) showed in a room with unclean sanitary conditions higher NH_3 (26.65 ppm vs. 18.17 ppm) and H_2S (0.099 ppm vs. 0.010 ppm) compared to a room with clean sanitary conditions. The deterioration of air quality can cause distress, drop in feed intake and predispose the respiratory tract to infectious agents.

3 Transition from weaning: intervention and prevention strategies

3.1 Creep feeding

In commercial swine production, creep feeding during the suckling period prior to weaning has been a common husbandry practice because it increases the weaning weight of piglets and leads to a smooth transition period for the piglets from sow's milk to the dry feed (Dong and Pluske, 2007; Cabrera et al., 2013). Previous studies demonstrated that creep feed intake during the suckling period has a positive effect on post-weaning feed intake (Bruininx et al., 2002; Cabrera et al., 2013), and it is assumed that nursery piglets offered creep feed prompts them to get adapted to solid feed (Dong and Pluske, 2007). This approach promotes gut development and therefore helps piglets more easily manage a dietary change after weaning (Jayaraman and Nyachoti, 2017). With respect to gut health, creep feeding helps to maintain nutrient supply after weaning, and consequently prevents villous atrophy, thereby reducing the chances of PWD in piglets (Pluske et al., 1996). Lallès et al. (2007) showed as *E. coli* infection is lower in piglets consuming an optimal amount of creep feed as compared with those consuming either no creep feed or high amounts before weaning. This finding illustrates the importance of voluntary feed intake regulation at this period.

3.2 Amino acids

Some traditionally classified dispensable amino acids, such as arginine, glutamine, glutamate and proline, play important roles in intestinal health. Supplementation of 0.4–0.8% L-arginine in a pre-weaning diet enhanced intestinal growth and development in the early post-weaning period (Yang et al., 2016). The administration of proline improved mucosal proliferation, intestinal morphology and tight junction and potassium channel protein expression in early-weaned piglets (Wang et al., 2015). Several mechanisms are highly involved in the benefits of glutamine or glutamine dipeptides on intestinal health (Wu et al., 2004, 2016; Wang et al., 2008). A growing evidence has revealed that the supplementation of indispensable amino acids, such as tryptophan and sulphur amino acids has positive effects on the intestinal health of weaned pigs by regulating host physiology, metabolism, oxidative status and immunity (Zong et al., 2018).

3.3 Phytochemicals and antimicrobial peptides

Dietary supplementation of phytochemicals, such as garlic, pepper, cinnamon, fennel, oregano, thyme, ginger, enhanced disease resistance (antibacterial and anti-viral effects, immune-regulatory activities and antioxidant action) and growth performance. These benefits were likely driven by improved gut health, such as improved intestinal barrier integrity (Zou et al., 2016).

Antimicrobial peptides, also known as host defence peptides, have been considered as potential alternatives to antibiotics in livestock (Robinson et al., 2018), possessing a strong and large-spectrum activity against gram-negative and gram-positive bacteria, fungi, parasites and viruses (Hancock, 2001). There are two ways to incorporate the benefits of antimicrobial peptides into animal health and nutrition. One is direct supplementation of exogenous antimicrobial peptides to animal feed, while the other one is to use dietary supplements/ingredients to stimulate the secretion of endogenous antimicrobial peptides by the host (Robinson et al., 2018). For example, Wang et al. (2006) reported that the supplementation of recombinant lactoferrin increased gut morphology and growth performance of piglets. Considering that the majority of exogenous antimicrobial peptides would be digested in the upper gastrointestinal tract, the stimulation of endogenous antimicrobial peptides secretion by nutritional manipulation may be a better approach (Xiong et al., 2019). Robinson et al. (2018) described the modulation of endogenous antimicrobial peptides synthesis in pigs by dietary compounds such as butyrate and vitamin D.

3.4 Organic acids and short-chain fatty acids

At weaning, natural acidification of the stomach via HCl secretion is reduced owing to immaturity of the digestive system and abrupt change in diet from milk to solid diets. Compared with the mature pig with a pH range of 2.0-3.0, the gastric pH in suckling and weanling piglets ranged between 2.6 and 5.0 (Heo et al., 2013). It is important to maintain a low gastric pH to optimise nutrient digestion and prevent pathogen overgrowth. Dietary addition of organic acids such as citric, fumaric, lactic and formic to weaned pig diets improved growth performance and health (Tsiloyiannis et al., 2001).

The SCFAs are a major fuel source for colonocytes and are essential for maintaining the normal metabolism of the colon mucosa. Other benefits of SCFAs can be summarised as below:

1 Modulate of the expression of genes involved in gut motility (Guilloteau et al., 2010),
2 Modulate the host defence (Guilloteau et al., 2010),
3 Modulate the inflammatory responses (Guilloteau et al., 2010),
4 Could stimulate the formation of intestinal barrier (Feng et al., 2018),
5 Protect intestinal barrier disruption (Feng et al., 2018),
6 May affect the composition of gut microbiota (Wen et al., 2012).

Among SCFAs, butyric acid has attracted significant research attention due to its importance in maintaining gut health in both humans and animals and has been shown to possess strong antimicrobial activity against both gram-positive and gram-negative pathogenic bacteria and effects on regulation of inflammatory responses and immunity (Xiong et al., 2019).

3.5 Probiotics and prebiotics

Probiotics are live microbial feed supplements. Commonly referred to as probiotics are *Bacillus*, yeast and lactic acid-producing bacteria such as *Lactobacillus*, *Bifidobacterium* and *Enterococcus* (Stein and Kil, 2006). Probiotics act inhibiting pathogen adhesion by steric hindrance or competitive exclusion, producing microbicidal substances such as bacteriosins and organic acids (Li et al., 2003) and increasing the production of SCFAs (Sakata et al., 2003).

Prebiotics, such as fructooligosaccharides and transgalactooligosaccharides, but also dietary fibre and resistant starch, are selectively fermented ingredients that allow specific changes, in both the composition and activity of microbiota, that confer benefits upon host well-being and health (Gibson et al., 2004; Heo et al., 2013).

3.6 Biosecurity

Academically, biosecurity can be defined as the application of measures aimed to reduce the probability of introduction and spread of pathogens (Barceló and Marco, 1988), to avoid transmission, either between farms or within the farm. A correct application of biosecurity measures implies a knowledge of the epidemiology of the diseases to be avoided, particularly of the routes of transmission, the stability of the agent in the environment and the role of fomites and vectors (Alarcón et al., 2021).

Biosecurity includes bio-exclusion, bio-containment and bio-management (Levis and Baker, 2011). Bio-exclusion (also known as external biosecurity) is preventing the entry of undesirable pathogens into the farm. Most pig producers concentrate on bio-exclusion as these types of practices are vital indicators for the chances of pathogen introduction, particularly in regards to the PRRSV (Bottoms et al., 2013).

Here are some very important rules that should be applied in all stages of production and in particular in the transition period from weaning.

Fomites carried by people (boots, clothes, etc.) or even the people itself, through contaminated skin, can spread various pig pathogens such as PRRSV, PEDV, TGEV and ETEC. Cleaning and disinfection of clothing, vehicles and equipment before and after contact with animals is essential. Boot baths should be provided at the entrance of the farm (Alarcón et al., 2021).

The control of visitors and fomites is a major focus, as both can bring pathogens to the farm (FAO, 2010). Visitors to farms should always be asked whether they have recently been to potentially contaminated places (FAO, 2010). The shower, before entering the clean area of the farm, should be compulsory. In the farm a delimitation between clean and dirty areas, perimeter fence and parking area outside the farm are important. The aim of these measures is to avoid vehicles delivering feedstuff, collecting dead animals or slurry entering the farm (Alarcón et al., 2021).

The transport of animals can be critical for the introduction in the farm of infectious agents. Trucks destined for the transport of replacements must not be used for the transport of animals to the slaughterhouse. The same truck should not pick up animals on different farms (Alarcón et al., 2021). The vehicles, especially those used to transport pigs, should be thoroughly cleaned before returning or visiting other farms (FAO, 2010). Bio-containment, avoiding the spread of pathogens from herd to herd, is mainly associated with external biosecurity.

Bio-management (also known as internal biosecurity) is characterised by the efforts to control infectious diseases already present in the farm population. Proper cleanliness and disinfection of the pig rooms, all-in/all-out and vaccinations, are the key components of internal biosecurity. In particular, the goal of all-in/all-out management system is to break the disease cycle by preventing the sharing of air-space of pigs carrying clinical disease with pigs susceptible to the infection by that disease (Jayaraman and Nyachoti, 2017). An important measure in preventing diseases in recently weaned pigs is the workflow, following the pig flow, from younger to older. Considering measures related to cleaning and disinfection, it is important to highlight that pens should be cleaned first by removing organic debris, then washed with soapy water, and after rinsing and drying they should be disinfected. As already reported for the external biosecurity, the personnel play a very important role in preventing the spread of the diseases among pigs. In particular, the respect of the assigned areas of work, the correct use of clothes, footwear, gloves and periodic hand washing and correct management of foot baths should be considered key factors (Alarcón et al., 2021). Assessing biosecurity includes measuring the potential routes for disease transmission. The most common biosecurity assessment has been the creation of scores. Most of these scores are based on values attributed to the biosecurity practices by expert opinion panels. Some of the scoring systems evaluate measures that are common to the transmission of different types of infectious agents while others are disease specific (Alarcón et al., 2021).

3.7 Vaccination

The aim of any vaccination is to prevent future disease, while, in many cases, it will not prevent future infection. For this reason, the main expected impact resides in a significant decrease of the clinical signs associated with the infection that the animals are vaccinated against. For several vaccines, a decrease in pathogen shedding in vaccinated animals has also been documented, with a reduction of the pathogen diffusion in the population (Rose and Andraud, 2017).

The transition from weaning is a challenging period from an immunological point of view. Maternal immunity tends to decrease from birth to weaning and, for many agents, it may no longer be present. At the same time, for some

Table 5 Examples of vaccines against the main pig diseases

Disease[a]	Target	Type of vaccine	Administration route
Swine erysipelas	• Piglets • Sows, gilts and boars	Inactivated or attenuated	Parenteral
Colibacillosis	• Piglets • Sows and gilts	Attenuated/Live non-pathogenic Inactivated	Oral Parenteral
Mycoplasmal pneumonia	• Piglets	Inactivated	Parenteral
Atrophic rhinitis	• Sows	Inactivated	Parenteral
Pleuropneumonia	• Pigs • Sows, gilts and boars	Inactivated: 1. killed organisms (bacterins); 2. subunit toxin-based vaccines; 3. bacterins + toxins	Parenteral
Ileitis	• Piglets	Attenuated Inactivated	Oral Parenteral
Glässer's disease	• Piglets • Sows and gilts	Commercial inactivated vaccines	Parenteral
Salmonellosis	• Piglets	Attenuated S. Choleraesuis/S. Typhimurium, alone or combined	Oral
Streptococcosis (*Streptococcus suis*)	• Piglets • Sows and gilts	Commercial inactivated vaccines	Parenteral
Parvovirosis	• Sows	Inactivated	Parenteral
PCVAD	• Pigs • Sows and gilts	Inactivated or subunit vaccines	Parenteral
PRRS	• Pigs • Sows, gilts and boars	Attenuated/Modified-live vaccine or inactivated	Parenteral
Porcine epidemic diarrhoea (PED)	• Sows and gilts	Inactivated or modified-live vaccine	Parenteral
Swine Influenza	• Sows and gilts	Inactivated	Parenteral
Rotavirosis	• Piglets • Sows and gilts	Attenuated/Modified-live vaccine	Oral/Parenteral
Aujeszky's disease	• Pigs • Sows, gilts and boars	Attenuated or inactivated	Parenteral

[a]Refer to the leaflets of the different vaccines for the vaccination scheme.

diseases, the period just before or after weaning is characterised by the need to vaccinate pigs. The timing of vaccination must take into account the onset of the disease, for which there is the need to vaccinate pigs and the possible interference of maternal immunity capable of interfering with the piglets' own immune response to vaccines. Vaccination of sows and gilts may influence the duration of maternal immunity and may interfere with vaccination efficacy in young piglets.

In Table 5, examples of pig and sow vaccination programmes have been reported, considering different strategies of protection involving pigs in the period of transition from weaning or soon after.

4 Weaning transition and infectious agents

The transition from weaning and the following 2-3 weeks represent a complicated health period in the pig's life. In this period, enteric, respiratory and systemic diseases are common. Multiple pathogens are often involved, and some of them can cause severe immunosuppression and severely worsen the incidence and severity of many diseases. This is the case of PRRSV that induces poor innate and adaptive immune responses (Done and Paton, 1995) and PCV2 viraemic pigs with an increased risk of secondary infections due to immune suppression or immunomodulation (Segalés et al., 2004; Shen et al., 2012).

Among agents causing enteric diseases, PWD caused by enterotoxigenic *E. coli* (ETEC) is one of the most important diarrhoeal diseases worldwide. ETEC PWD is also an example of a disease influenced by predisposing risk factors associated with the transition from weaning. In this period, several agents can be involved in the aetiology of the porcine respiratory disease complex (Opriessnig et al., 2011), with huge differences between countries, regions and farms. Typical of the transition from weaning and the weeks soon after this period is the systemic forms caused by *Streptococcus suis* and *Glaesserella parasuis*, recognising predisposing factors such as concomitant infections, poor hygiene, stress and low biosecurity measures.

5 Viruses acting in the period of transition from weaning as immunosuppressive agents

5.1 Porcine reproductive and respiratory syndrome virus (PRRSV)

5.1.1 Impact and epidemiological characteristics

The PRRS virus causes viraemia, which typically leads to pyrexia, pneumonia with abnormal respiratory behaviour and reduced average daily weight gain (ADWG) (Martelli et al., 2009). Numerous studies have shown that pigs similarly infected with PRRSV elicit diverse response profiles. While some manage to clear the virus within a few weeks, others experience a prolonged infection with a rebound (Go et al., 2019). In an experimental study, Islam et al. (2013) demonstrated that around 20% of pigs exhibited viraemia rebound within 6 weeks postinfection, confirming that this phenomenon is common and could be due to a genetic factor, that would lead to variable immune responses (Boddicker et al., 2012). PRRSV seroconversion and viraemia typically occur during the nursery phase of production. It is a common practice for veterinarians and producers to adjust the timing (delay or move earlier) of the administration of common vaccinations (Drew, 2000) to avoid vaccine failure associated with PRRSV-induced immunosuppression. Overall, the immune responses against PRRSV are ineffective in resolving the infection completely, resulting in prolonged viraemia and persistent infection in lung and lymphoid tissues, potentiating the effects of other swine pathogens (Murtaugh et al., 2002).

It is important to know the main epidemiological characteristics of PRRSV in order to tackle the effect of PRRSV infection in pigs, implementing appropriate measures of control (Table 6).

5.1.2 Control measures

A key aspect to any control programme is the management of the gilt pool (Dee et al., 1994). The introduction of naïve replacement gilts into an infected breeding population results in recurrent episodes of reproductive failure and both horizontal and vertical transmission of the virus from the dam to offspring. To properly acclimatise incoming replacement gilts, many production systems use designated facilities for gilt development units (Dee et al., 1997). Replacement gilts enter these facilities at a young age (at weaning or at 25 kg/BW) and this allows them adequate time (4-6 months) for the development of protective immunity through the use of modified-live vaccines and/or exposure to farm-specific wild-type virus.

The selection of efficacious products and protocols for sanitising livestock transport vehicles and other surfaces in the farms when PRRSV is involved is extremely important in its control. PRRSV is inactivated by lipid solvents, for example chloroform and ether, and it is highly unstable in solutions containing low concentrations of detergents, which disrupt the envelope with concomitant release of the non-infectious core particles and loss of infectivity (Snijder and Meulenberg, 2001). At 'room temperature', complete inactivation of the virus is accomplished with chlorine (0.03%) in 10 min, iodine (0.0075%) in 1 min and a quaternary ammonium compound (0.0063%) in 1 min (Shirai et al., 2000). Decontamination protocols involving drying, thermo-assisted drying and foaming disinfectants containing glutaraldehyde and quaternary ammonium

Table 6 The main epidemiological characteristics of PRRSV influencing risk factors and measures of control

Main topics	Main subcategory	Rationale/specifications
Stability in the environment	PRRSV thermal stability and disinfection	Inactivation[a]: • solvents, heat, drying, or at a pH below 5 or above 7 • iodine (0.0075%) • quaternary ammonium compounds (0.0063%) • chlorine (0.03%) PRRSV is stable at: • −70°C and −20°C (months or years)
Routes of shedding	• Oral and nasal secretion • Semen • Urine • Faeces • Mammary secretion	Shedding of PRRSV in semen reported for up to 101 days postinfection[b]
Persistent infection	The most significant epidemiological feature of PRRSV infection	Detection of infectious virus for 100–165 DPI[b]
Transmission	Route of transmission: • Intranasal • Intramuscular • Oral • Intrauterine • Vaginal Indirect transmission: • equipment • instruments • clothing • water • food • aerosols • arthropod vectors	Dose required to infect one-half of the exposed animals[b]: • Intranasal: $1 \times 10^{5.3}$ TCID$_{50}$ • Oral: $1 \times 10^{4.0}$ TCID$_{50}$ • Exposure via artificial insemination: $1 \times 10^{4.5}$ TCID$_{50}$ Meteorological conditions supportive of airborne transport and transmission[b]: • low temperatures • moderate levels of relative humidity • rising barometric pressure slow directional winds
Vertical transmission	• Foetal death • Birth of weak infected pigs • Birth of infected pigs that may appear normal	PRRSVs cross the placenta efficiently only in the last trimester of pregnancy[a]
Introduction in a herd and transmission between herds	• Infected pigs • Virus-contaminated semen • Aerosols in herd-to-herd transmission • Lack of biosecurity measures • Infected transports • Insects • Herd density • Proximity to infected herds • Personnel • Equipment • Wild boar	• Airborne transport of PRRSV reported out to 4.7 km[c] • Wild boars: Infection confirmed in different European countries[d] • Houseflies: mechanically harbour the virus for up to 48 hours. PRRSV replication not supported[e]

[a] Pileri and Mateu (2017).
[b] Zimmerman et al. (2019).
[c] Dee et al. (2009).
[d] Bonilauri et al. (2006); Reiner et al. (2009); Vilcek et al. (2015); Roic et al. (2012); Albina et al. (2000); Oslage et al. (1994); Baroch et al. (2015).
[e] Schurrer et al. (2005).

chloride compounds are effective at inactivating PRRSV in farrowing rooms and transport vehicles in cold and warm climates (Dee et al., 2004). PRRSV ultraviolet light exposure (10 min) completely inactivated the virus on common farm surfaces and materials (Dee et al., 2011).

Treatment of vehicles by washing at high temperature (80°C), followed by phenol disinfection and overnight drying, was effective for complete sanitation of trailers. Alternatively, the use of a thermo-assisted drying and decontamination system or glutaraldehyde fumigation had equivalent efficacy to overnight drying for the complete trailer decontamination (Dee et al., 2005).

5.1.3 Vaccination

Vaccination of sows and piglets is one of the strategies commonly used for controlling PRRS; however, none of the vaccines commercially available has been fully effective in preventing the spread of the virus within a herd (Shirai et al., 2000; Pileri and Mateu, 2016) (Fig. 3). The most common PRRS vaccines can be broadly categorised into modified-live virus (MLV) vaccines or inactivated (IV) vaccines. PRRS MLV vaccines are attenuated live vaccines, which have shown effective protection against homologous and partial protection against heterologous PRRSV strains (Shirai et al., 2000; Chae, 2021), reducing clinical signs, severity and the duration of viraemia and virus shedding (Pileri and Mateu, 2016).

MLVs are able to replicate in the host, to develop viraemia for up to 4 weeks after immunisation, leading to the spread of vaccine virus to naïve animals (Charerntantanakul, 2012) and inducing an immune response similar

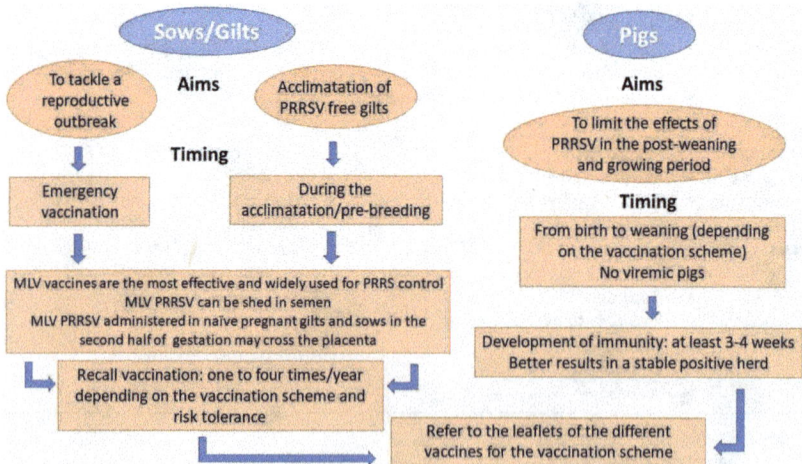

Figure 3 General considerations on PRRSV vaccination strategies (Zimmerman et al., 2019; Chae, 2021).

to that induced by mildly virulent PRRSV isolates. As mentioned earlier, the virological and clinical protection afforded by MLV vaccination is considered partial against a heterologous PRRSV strain. However, in this case, vaccinated pigs can show a viraemia of shorter duration compared to non-vaccinated pigs (Charerntantanakul, 2012; Roca et al., 2012).

The limited protective immunity and immunogenicity to many circulating PRRSV strains have raised major concerns regarding vaccine effectiveness (Shirai et al., 2000; Klinge et al., 2009). The potential efficacy of MLV vaccines in the field can be estimated:

- indirectly, by assessing the biological parameters related to transmission (i.e. duration of viraemia and shedding of the virus, number of chronically infected pigs, etc.),
- directly, by determining the reproduction rate of PRRSV (Pileri and Mateu, 2016) as most vaccines reduce pathogen transmission not only by offering protective immunity but also by reducing host infectivity and duration of infection (Bitsouni et al., 2019).

In this way, MLV vaccines, despite being far from perfect, could achieve drastic reduction in the occurrence and severity of PRRS outbreaks in commercial pig populations, and even help to eliminate the disease (Bitsouni et al., 2019).

PRRS inactivated vaccines show high safety, but the protective immunity that they provide against either homologous or heterologous PRRSV strains is extremely limited, and they often fail to significantly reduce clinical signs, viraemia and duration of shedding in naïve animals (Christianson et al., 1993; Albina et al., 1994). Consequently, MLV vaccines are the predominating vaccines in the field nowadays (Chae, 2021).

5.2 Porcine Circovirus type 2 (PCV2)

5.2.1 Impact and epidemiological characteristics

The viraemia is a characteristic feature of PCV2 infection. In conventional farms, viraemic animals are often present during the nursery and fattening stage. It has been seen that the presence of viraemic animals due to PCV2 occurs with a higher frequency and earlier in farms with PCV2 systemic disease. This condition is very rarely observed in pigs younger than 4 weeks of age (Segalés and Domingo, 2002), suggesting that certain levels of maternal immunity protect against the development of PCV2 systemic disease (Calsamiglia et al., 2007). Following PCV2 infection an important proportion of the animals can show intermittent and/or long-lasting viraemia that can go on until the 28th week of age (Rodríguez-Arrioja et al., 2002). A high PCV2 viraemia has already been shown to be associated with the development of porcine circovirus-associated diseases (Olvera et al., 2004; Meerts et al., 2006). Lopez-Soria et al. (2014) showed that the amount of PCV2 in blood had a negative effect on the ADWG, reporting that the higher the PCV2 load the lower the ADWG.

5.2.2 Vaccination

The most effective tool to tackle the production losses related to the PCV2 infection is to induce a passive or an active immunity that minimises the circulation of the virus in the farm. Most current vaccines worldwide are based on PCV2a strains (Segalés, 2015) and their efficacy against mixed infection of PCV2a and PCV2b and the recently emerged PCV2d is not so clear (Opriessnig et al., 2014; Hou et al., 2019). Different strategies of vaccination are available in sows and piglets (Fig. 4): (1) vaccination of sows for offspring and/or reproductive disease protection; (2) piglet vaccination as the way to control porcine circovirus-associated diseases in the farm and (3) vaccination in both sows and piglets. In this last approach, it is important to consider the possible interference of maternally derived antibodies (MDA) (Segalés, 2020). Martelli et al. (2016) concluded that some interference of MDA with the induction of an efficient immune response could be considered in case of vaccination of sows and piglets, even if, in a condition of high levels of MDA, piglets vaccination at 4, 6 or 8 weeks of age confers a protective immune response with a long-lasting (until 34 weeks of age) antibody response.

The reduction of PCV2 viraemia by the vaccine plays a critical role in controlling PCV2 infection (Chae, 2012) reducing significantly the percentage of viraemic piglets, as well as the viral load in the blood and the virus excretion (Seo et al., 2014).

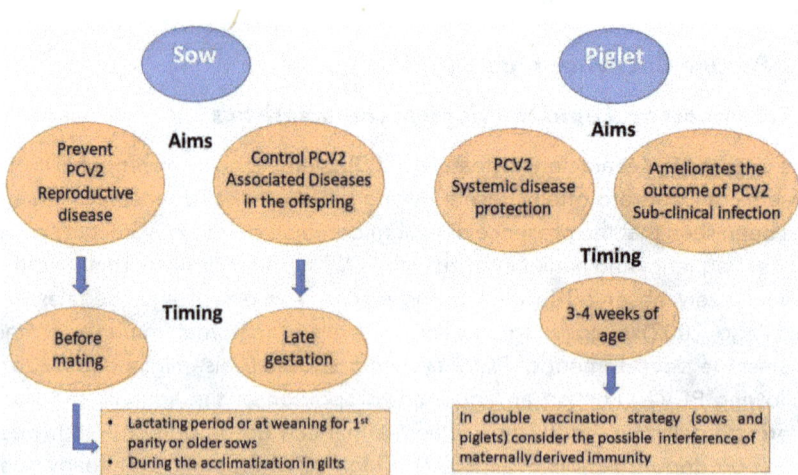

Figure 4 PCV2 vaccination strategies (Segalés, 2020).

Vaccination failure has been noticed in the last few years, showing an increase of PCV2 systemic disease in farms with vaccinated piglets. In these cases probably, the vaccination at weaning might not provide a sufficient vaccine-elicited immune response before natural infection occurs and a proportion of animals may develop PCV2 systemic disease and not just a PCV2 sub-clinical infection (Segalés, 2020). This condition is probably mostly associated with an inadequate management of the vaccine and timing of application (Segalés, 2020):

- too early, with potential interference with maternally derived immunity,
- too late, in case of early infections (too close to natural infection),
- in diseased animals (for example in PRRSV viraemic pigs).

6 Example of diseases recognising risk factors and prevention strategies in the transition from weaning

A large number of infectious agents are capable of causing disease in young weaners. *E. coli* is the major cause of PWD, enterotoxemia and mortality. Other infectious agents can determine severe episodes of diarrhoea in recently weaned pigs (Table 7), in some cases, recognising similar risk factors and measures of control.

6.1 Enterotoxigenic E. coli post-weaning diarrhoea: a multifactorial disease

The most important cause of PWD is associated with the proliferation of enterotoxigenic *E. coli* (ETEC) in the small intestine of pigs, in particular those

Table 7 Etiological agents of post-weaning diarrhoea (modified from Martelli, 2013)

Disease/etiological agent	Age	Lethality
Colibacillosis *Escherichia coli* (ETEC, EPEC)	Most commonly post-weaning until 45-50 days	Can reach 25%
Swine dysentery *Brachyspira hyodysenteriae*	6 weeks old to adult; frequent in the growing-fattening periods	Variable, usually low
Salmonellosis (*Salmonella* Typhimurium)	All ages, usually after weaning, mostly in the growing-fattening periods	Low
PED and TGE Coronavirus PEDV TGEV	All	Can be high; less severe than in neonates
Rotaviral enteritis Rotavirus	From 1 to 5 weeks	Low, < 20%
Proliferative enteropathy *Lawsonia intracellularis*	Post-weaning (5 weeks old to young adults)	Low

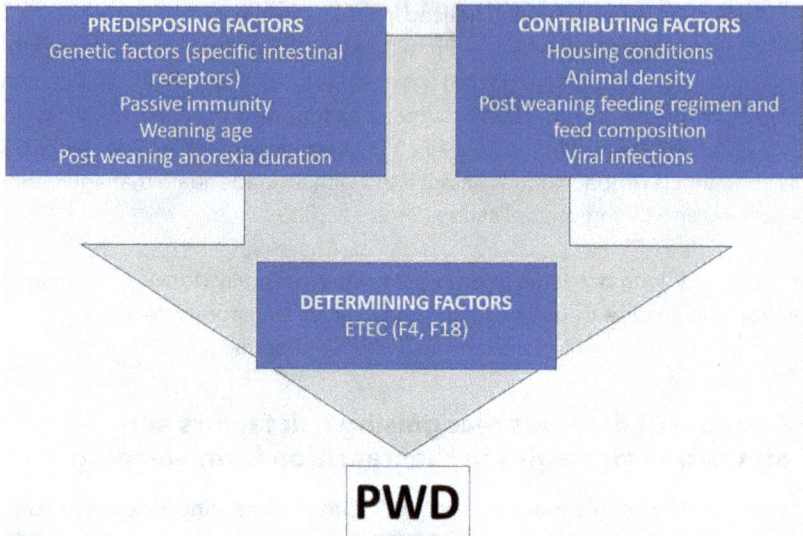

Figure 5 The multifactorial genesis of PWD in pig involves interaction between predisposing, contributing and determining factors (modified from Rhouma et al., 2017).

that express fimbrial adhesins F4 (K88) or F18 (determining factors). However, ETEC PWD shows a multifactorial genesis involving the interaction between predisposing, contributing and determining factors (Rhouma et al., 2017) (Fig. 5). Predisposing factors are elements that increase the risk of developing a disease (Racine et al., 2016). A contributing factor is a condition that influences the effect and severity of the infection. In other terms, outbreak contributing factors are conditions that enable or amplify an outbreak (Brown et al., 2017). The determining factors are considered as the causal factors of post-weaning colibacillosis such as ETEC (Rhouma et al., 2017).

6.1.1 Enterotoxigenic E. coli post-weaning diarrhoea: predisposing factors

The pig's susceptibility to developing an ETEC F4 and F18 infection depends on the presence and amount of F4 and F18 receptors, respectively, found in the brush border membrane of the small intestine, allowing the ETEC to adhere and develop diarrhoea (Jørgensen et al., 2003). The locus encoding the intestinal receptor responsible for adhesion of F4 ETEC was mapped to the q41 region on pig chromosome 13 (SSC13) (Fontanesi et al., 2012) and among the candidate genes found in this region (MUC4, MUC13, TFRC and MUC20), the MUC13 and MUC4 are the most likely responsible genes (Zhang et al., 2008). In addition tyrosine kinase non-receptor 2 (ACK1) and UDP-GlcNAc:betaGal beta-1,3-*N*-acetylglucosaminyltransferase 5 (B3GNT5) genes have been proposed as genetic markers for pig ETEC resistance/susceptibility (Wang et al., 2007; Ouyang et al., 2012). Regarding pig resistance to ETEC F18 infection, two main single nucleotide polymorphisms (SNPs) localised on alpha (1,2)-fucosyltransferase (FUT1) and bactericidal/permeability-increasing protein (BPI) (Meijerink et al., 2000; Liu et al., 2013) genes, respectively, have been proposed and greater consensus has been attained for the SNP located on FUT1.

Piglets are born with an immature mucosal immune system, which develops over the first few weeks of life following a programmed sequence, without reaching a complete maturity at weaning ages which are common on commercial farms (Stokes, 2017). At weaning the piglet's mucosal immune system is required to distinguish between antigens and recognise and respond appropriately to potential pathogens (Stokes, 2017). The relative immaturity of the intestinal immune system at weaning age (characterised by the lack of a previous stimulation by pathogens) may reduce the ability of the weaned pig to mount an appropriate immunological response to pathogens (Heo et al., 2013). This condition is coupled with the sow milk removal, and consequent discontinuation of nutritive intake of the IgA present in this milk contributes to increase the susceptibility of pigs to microbial infections (Rhouma et al., 2017).

Weaning age is an important ETEC PWD predisposing factor. Weaning pigs at an early age, before 21 days, increase the occurrence of PWD, encouraging them to move away from this practice and as a result wean pigs no earlier than 26 days of age (Rhouma et al., 2017).

Feed intake is usually reduced initially after weaning and this is strongly correlated with the risk of disease occurrence over the post-weaning period (Pluske et al., 1997; Vente-Spreeuwenberg et al., 2003). This condition promotes the creation of an ideal environment for the multiplication of bacteria such as pathogenic *E. coli* and allows toxins and bacteria to cross the epithelium as a result of intestinal inflammation (Campbell et al., 2013).

6.1.2 Enterotoxigenic E. coli post-weaning diarrhoea: contributing factors

Among the contributing factors, housing conditions are very important; in particular, it is essential to provide the correct environmental temperature, 26–28°C, to maintain pigs in their thermo-neutral zone, avoiding drafts and removing moisture and gases using adequate ventilation (Rhouma et al., 2017). Automatic temperature control was associated with a decreased risk of PWD (Laine et al., 2008).

Mixing piglets from different farms is a common practice in pig husbandry, resulting in fighting with aggressive interactions, occurring typically during the first few hours after grouping (Coutellier et al., 2007).

Post-weaning feeding regimen and feed composition play a role in the development of ETEC PWD. In Fig. 6, have been reported the effects of different nutrients on the incidence of PWD.

Resistant starch improves colon health thanks to the fermentation in the large intestine by colonic microflora and then is digested into hydrogen, methane and SCFAs, such as acetic, propionic and butyric acid (Alsaffar, 2010) (Fig. 7). The higher proportion of resistant starch showed a reduction in ileal and caecal digesta pH (Gao et al., 2019) improving the resistance to infective diseases (Mateos et al., 2007).

Diets characterised by high crude fibre (10-17%) and low in nutrients, particularly CP reduce proliferation and excretion of pathogenic *E. coli*. Considering that the apparent digestibility of CP at the terminal ileum is 75-85%, diets containing 220-250 g CP/kg, result in a considerable amount of protein entering the large bowel. As an example, a 7 kg pig eating 300 g/day of a 220 g CP/kg diet could lose 33-55 g protein daily to its hindgut (Pluske et al., 2002). Interestingly the activity of major gut proteases (trypsin and chymotrypsin) was stimulated after weaning by an increase in the level of CP in the weaning diet not exceeding 200 g/kg. This would explain data showing a decrease in the apparent ileal digestibility of nitrogen in pigs fed diets containing more than 225 g CP/kg (Li et al., 1993). Protein is fermented rapidly by the microbiota with the production of biogenic amines such as putrescine, cadaverine and tryptamine and gases (e.g. NH_3) that have been implicated in the clinical expression of ETEC PWD (Pluske et al., 2002).

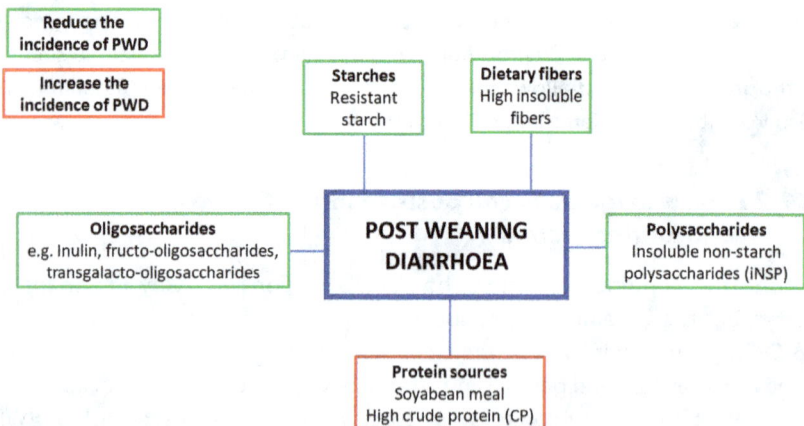

Figure 6 Effects of different nutrients on the incidence of PWD (Gerritsen et al., 2012; Rhouma et al., 2017; Liu et al., 2018; Gao et al., 2019).

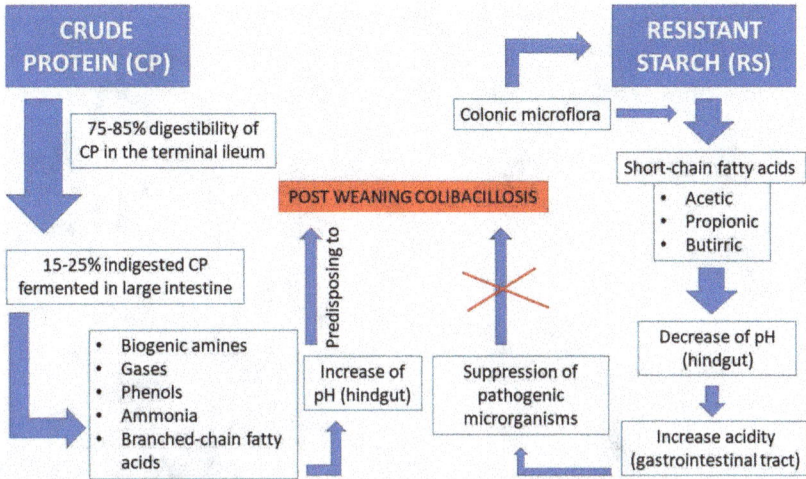

Figure 7 Role of crude protein and resistant starch in predisposing and preventing pig to PWD (Gao et al., 2019).

Apart from the feed's composition, the management of its administration also plays an important role. Interestingly it was shown that the prevalence of PWD was higher on farms that fed weaned piglets only twice a day with a restricted amount of feed, than on farms that provided more than two meals per day with or without feed restriction (Laine et al., 2008).

Considering the effect of other pathogens and in particular viral agents, as contributing factors, the isolation of ETEC strains in weaned pigs can often be accompanied by rotavirus or by the coronavirus of porcine epidemic diarrhoea (PED) (Nagy et al., 1996). Rotavirus is an important etiological agent and contributing factor of PWD, with different prevalence. In the USA, Canada and Mexico, rotavirus type A was detected in diarrhoeic pigs with the prevalence of 84% and 61%, from 21 to 55 days and >55 days old pigs respectively (Marthaler et al., 2014), while in Italy the 71.5% of pigs suffering from diarrhoea, from 28 to 84 days old, were positive for rotavirus type A (Martella et al., 2007). Co-infection with rotavirus and ETEC results in more severe disease, compared to the outcome of the infection with a single etiological agent (Tzipori et al., 1980). PRRSv infection can predispose pigs to ETEC infection, enteric colibacillosis and *E. coli* septicaemia mainly through an impairment of the immune response of piglets (Rhouma et al., 2017).

6.1.3 Strategies to control enterotoxigenic E. coli post-weaning diarrhoea

Several strategies can be implemented to prevent and control ETEC PWD, mainly based on feeding strategies, as reported above (see also Chapter 2), use

Table 8 Benefits and limitations of the major preventive strategies for the control of ETEC PWD in pigs (modified from Rhouma et al., 2017)

Strategies	Benefit	Rationale	Strategy limitations
Achieve high post-weaning feed intake/increase weaning age	Reduce the risk for PWD	Minimise post-weaning anorexia	Economically not convenient
Avoid some ingredients (such as soybean meal)	Reduces the severity and frequency of PWD	Soybean meal reduces: • duodenal specific activities intestinal enzymes • intestinal digestibility. Contradictory results have been reported	Growth retardation
Feeds with decreased protein content	Decrease *Escherichia coli* colonisation and to minimise PWD prevalence	High protein diet cause protein maldigestion	Growth retardation
Stocking density reduction	Decrease the occurrence of PWD and other diseases	Reduce stress factors	Significant costs
Diagnostic tools and appropriate diagnostic path	Avoid the inappropriate use of antimicrobials and improve the prevention	Correct diagnostic criteria to confirm ETEC PWD	Need of specialised laboratories
Vaccination	Specific protection against ETEC F4 and F18	Commercial F4 and F18 live avirulent vaccines are available	Interference with the lactogenic immunity
Selection of pigs genetically resistant to F4 and/or F18	Very effective approach	Pigs resistant to ETEC F4 and/or F18 do not express intestinal receptors for these fimbrial types	Expensive process Lack of techniques for a large-scale selection
Zinc Oxide (ZnO) administration	• Inhibition of bacterial adhesion to the intestinal mucosa • Growth rate stimulation • Integrity of intestinal mucosa • Modulate immune functions	ZnO at the levels of 2400-3000 ppm in pig feed was effective in reducing PWD	• Soil heavy metal contamination • Bacterial resistance • Co-resistance • No longer available from June 2022

Strategies	Benefit	Rationale	Strategy limitations
Addition of organic acid to reduce gastric pH	• Modulate immune function • Lower pH in the stomach • Improve growth performance • Reduce PWD	Exposure to low pH values (i.e. 3.0–4.0) is bactericidal for many pathogenic bacteria	Exact modes of action are still unknown Antimicrobial activities is different between acids
Addition of prebiotics and probiotics	• Improve intestinal health • Improve growth performance • Reduce ETEC attachment to intestinal mucosa • Reduced diarrhoea	Different mechanisms: • Inhibition of bacterial toxins production • Inhibition of the adhesion of pathogens to the intestinal mucosa	Sometimes contradictory studies and results on their effectiveness
Dehydrated porcine plasma administration	Improve growth performance Reduce incidence and severity of diarrhoea	Reduction of inflammatory cytokines expression Maintenance of mucosal integrity, and enhancement of specific antibody defense	High cost Rigorous control during the preparation process

of antimicrobials, prevention of predisposing viral infections, ETEC vaccination and use of zinc oxide (ZnO) (Table 8).

Antimicrobial treatments have been used administered in feed or water to treat and to prevent ETEC PWD (Luppi, 2017). Nowadays a prophylactic use of antimicrobials is no longer acceptable and preventive husbandry, active immunoprophylaxis and dietary preventive measures are the main tools for controlling PWD in weaned pigs.

To protect newly weaned piglets, an active intestinal mucosal immune response, in which the production of antigen-specific secretory IgA plays an important role, is required (McGhee et al., 1992). By vaccinating the oral route is the most logical route to obtain this type of response (Melkebeek et al., 2013). Several oral vaccines have been successfully implemented in weaned pigs, including subunit vaccines as well as live oral vaccines. A bivalent F4/F18 live vaccine comprising a F4 and a F18 non-pathogenic strain has been recently licensed in several countries for immunisation of pigs against post-weaning enteric colibacillosis (Fairbrother and Nadeau, 2019).

The addition of zinc oxide at the levels of 2400–3000 ppm in pig feed was demonstrated to reduce the incidence of PWD and mortality, improving growth performance in weaned pigs (Roselli et al., 2003). On 26 June 2017, the European Commission adopted a decision to withdraw all marketing authorisations for veterinary medicinal products, listed in Annex I of the decision, containing zinc oxide administered orally to food-producing species. The Member States have had the possibility to defer the withdrawal of the marketing authorisations for up to five years from this date, so the decision must be implemented no later than 26 June 2022.

6.2 Streptococcosis

6.2.1 Epidemiology and predisposing factors

S. suis is a major porcine pathogen causing high morbidity worldwide, including well-managed herds with high hygiene standards. Transmission of virulent strains between herds usually occurs by the movement of healthy carrier animals. The introduction of carrier pigs harbouring virulent strains (breeding gilts, boars and weaners) into a noninfected recipient herd may result in the subsequent onset of disease in weaners and/or growing pigs. Sows infect their piglets during farrowing via contamination from vaginal colonisation and probably through the respiratory route (Segura et al., 2016). Even though different serotypes and strains within the same serotype are present in a herd, a single strain usually causes most diseases (Marois et al., 2007). Horizontal transmission is important, especially during outbreaks, when diseased animals shed higher numbers of bacteria, thereby increasing transmission by direct contact or aerosol (Cloutier et al., 2003). The oral route

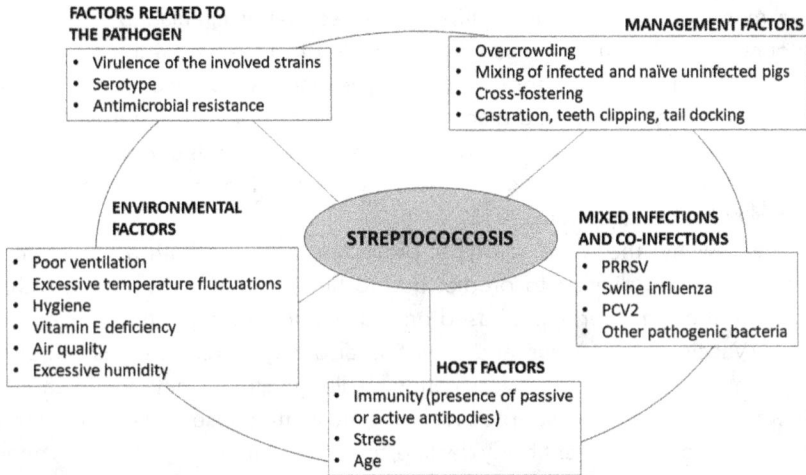

Figure 8 Main factors influencing and modulating the development of streptococcosis in pigs (modified from Obradovic et al., 2021).

of infection is not yet been proven, and recent studies suggested that *S. suis* is not able to survive in feed pellet or crumb feed, with or without formic acid or in the stomach contents (Warneboldt et al., 2016). *S. suis* may also be transmitted via fomites (Dee and Corey, 1993) and flies (Enright et al., 1987), although the importance of such vectors has still to be confirmed. In addition to the virulence of involved strains, other factors can impact the development of the disease such as overcrowding, poor ventilation and climatic conditions, poor hygiene status, high air pollution load and other stressors (Vötsch et al., 2018) (Fig. 8). Host-specific factors, such as age, genetic background and immunosuppression, influence the disease's development. Weaning piglets are most susceptible since protective maternal antibodies decline (Cloutier et al., 2003).

6.2.2 Control and preventive measures

Beta-lactam antibiotics (penicillin, ampicillin, amoxicillin and ceftiofur), gentamicin, tiamulin and a combination of trimethoprim and sulphonamide are the most useful antibacterial products for parenteral treatment (Gottschalk and Segura, 2019). Vaccines used in the field to prevent *S. suis* disease are either autogenous or (a very few) commercial bacterins, and results have been inconsistent (Segura, 2015). Possible reasons for the failure of autogenous vaccines include degradation of protective antigens caused by heat or formalin processing, inadequate bacterial antigen concentrations, the adjuvant included in the formulation, low production of antibodies and/or production of antibodies to antigens not associated with protection (Segura, 2015;

Rieckmann et al., 2020) and multiple *S. suis* serotypes circulating in the farm (Segura et al., 2020). For these reasons, it is of fundamental importance to obtain clear information about the serotypes/strains involved in the clinical cases, following appropriate diagnostic procedures (Fig. 9). In herds infected with multiple strains or serotypes of *S. suis*, multivalent vaccines or vaccines that provide a strong degree of cross-immunity are needed to provide adequate control of infection (Segura, 2015).

Even within the same serotype, protection obtained with a commercial bacterin may be difficult to predict due to high phenotypic variation within strains of the same serotype, based on available data for serotype 2 (Segura, 2015). Vaccination of sows and gilts has also been described as effective (Swildens et al., 2007). Most vaccination-challenge studies have been carried out with piglets. Because streptococcosis is most often observed at 6-10 weeks of age, due to maternal antibody decline, the first of two doses of bacterin need to be given at approximately 3-4 weeks of age and interference by maternal antibodies must be considered (Lapointe et al., 2002). Sow vaccination is less costly and labour intensive, thus representing an economical alternative to piglet vaccination. However, without subsequent active vaccination of the piglets to lengthen protection, this strategy could lead to highly susceptible growers (Baums et al., 2010). Preventing measures based on feed additives showed encouraging results. Correa-Fiz et al. (2020), in a comparative study about the prevention of streptococcosis, tested the effects of feed supplementation with medium-chain fatty acids and a natural anti-inflammatory. In this study clinical signs compatible with *S. suis* obtained with this supplementation were at the same level as those obtained with amoxicillin, concluding that feed additives may constitute an alternative to metaphylaxis.

6.3 Glässer disease

6.3.1 Epidemiology and predisposing factors

G. parasuis is a member of the normal respiratory microbiota, ubiquitous in swine herds worldwide, colonising the upper respiratory tract of piglets soon after birth through contact with the sow, which is the main reservoir and transmitter of *G. parasuis* to the piglets. Glässer's disease is commonly developed after weaning due to a reduction in maternal antibodies, mix of litters, strains variations, and other stress factors (Costa-Hurtado and Aragon, 2013; Costa-Hurtado et al., 2020). Basically, all pig farms are positive to *G. parasuis*, but not all farms develop the disease (Mahmmod et al., 2020) and this is usually due to specific strains, sometimes following a predisposition of the host due to distress factors and/or co-infections (Aragon et al., 2012), such as PRRSV and PCV2 (Pereira et al., 2017).

Among factors affecting Glässer's disease epidemiology, Mahmmod et al. (2020) showed as multi-site production type increases the risk of Glässer's disease in comparison to the one-site production type (farrow-to-finish). This could be due to stress factors caused by animal handling, transportation and relocation from one environment and management conditions to another. For these reasons, efforts should be done to achieve management practices that minimise stressful situations, particularly on the weaning phase (Costa-Hurtado et al., 2020).

Correa-Fiz et al. (2016) reported that the nasal microbiota composition was associated with the clinical status of the farm of origin of the piglets, leading to different susceptibilities to invasive infection by *G. parasuis*. Some microbial populations, such as *Chitinophagaceae* and *Corynebacteriaceae* families, may facilitate the colonisation of virulent *G. parasuis* strains, which are inhibited by *Enterobacteriaceae* and *Peptostreptococcaceae*, meanwhile, members from the *Flavobacteriaceae* family could facilitate the colonisation of non-virulent *G. parasuis* strains (Mahmmod et al., 2020). The interactions reported could be a base for innovative non-antimicrobial alternatives for Glässer's disease control.

6.3.2 Control and preventive measures

Antibiotics are widely used to prevent and control the disease, but increasing pressure to reduce antibiotics usage, particularly for prophylaxis strategies, puts more emphasis on vaccination strategies to prevent systemic infection and mortality.

The early colonisation of pigs with virulent strains in the presence of maternal immunity actually controls the evolution of the infection, preventing the disease and mortality post-weaning (Oliveira et al., 2004), avoiding subsequent Glässer's disease in the farms (Correa-Fiz et al., 2016). However, contradictory results have been obtained on the influence of MDA on nasal colonisation of piglets (Cerdà-Cuéllar et al., 2010; Brean et al., 2016). Sow vaccination may be effective to protect piglets during lactation, but lasting protection may require vaccination of piglets after the farrowing phase, together with actions to ensure early colonisation. There is not a complete agreement on the interference of the MDA with piglet vaccination against *G. parasuis*. While some studies report no effective interference when vaccinating piglets with two doses (Oh et al., 2013), other studies report an interference effect after sow vaccination that precluded the effectiveness of vaccination in piglets (Pomorska-Mol et al., 2011).

Commercially available bacterin vaccines are based on serovar 5, combined serovars 4 and 5, or combined serovars 1 and 6 for use in sows and pigs (Zhao et al., 2017). All these products show variable cross-protection to heterologous strains (Aragon et al., 2019). When a commercial vaccine fails

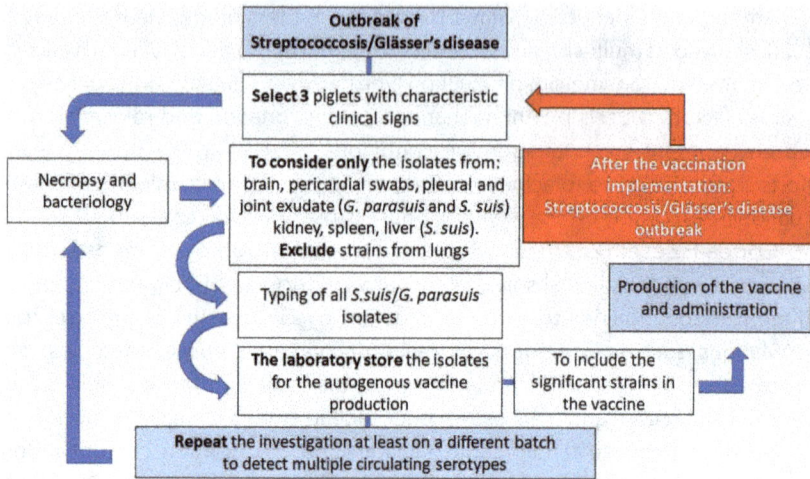

Figure 9 Steps and criteria for selecting strains of *Streptococcus suis* or *Glaesserella parasuis* for inclusion in an autogenous vaccine.

to protect vaccinated herds, the most appropriate solution, in a short term, is to develop autogenous vaccines, which need to be developed rationally. It is recommended to include in the formulation, strains necessarily typed and characterised, isolated from systemic sites such as meninges, pericardium and joints (Oliveira and Pijoan, 2004; McOrist et al., 2009), avoiding isolates from the respiratory tract (Fig. 9).

7 Conclusion and future trends

The weaning transition period, as described in this chapter, is a very stressful phase in pig's life in which several factors act in a synergistic way, predisposing piglets to poor health and reducing their performance.

Measures aimed at improving animal health and welfare in the period of transition from weaning, are of huge importance. Among these the stocking densities reduction, the improvement of piglet health at weaning, the diagnostic approach, and the vaccination strategies implementation are the most important. A very important challenging point is that veterinarians and farmers update the concept of control of the diseases intended as a control of a single disease, switching it in the control of an entire system, where we have complex interaction between pathogens and predisposing factors.

The next steps to improve the health of recently weaned pigs should focus the research on four main fields. The first is about the improvement of knowledge about the diagnostic approach, in order to reach the correct diagnosis and consequently correct measures of prevention and control. Even if this may seem a sufficiently developed topic, it is frequent to have in the practice, diagnostic conclusions not supported by correct criteria. The second concerns the study of the impact of pig's genetics on disease resistances and how this aspect modifies the epidemiology of the diseases themselves. The third aspect concerns the study of the microbiota and its impact on diseases, as well as the tools available to influence the bacterial populations that compose it. Last but not least, is the development and use of technology to improve the health and performance of pigs. The implementation of sensors to monitor parameters such as humidity, temperature, ammonia, CO_2 and dust are of fundamental importance to check environmental variables and their impact on pig health. Cameras to detect problems such as lameness or immobile animals, to identify aggressive behaviours such as fights or tail bites can be useful tools, as well as devices for measuring the animal temperature and individual feed consumption. All the data collected using these tools can be extremely valuable, but only if taken accurately, analysed correctly and presented in a way that allows veterinarians to make decisions.

8 Where to look for further information

8.1 Further reading

The following articles provide a good overview of the subject:

- Alarcón, L. V., Allepuz, A. and Mateu, E. (2021). 'Biosecurity in pig farms: a review', *Porcine Health Management* 7:5. https://doi.org/10.1186/s40813 -020-00181-z.
- Drew, T. W. (2000). A review of evidence for immunosuppression due to porcine reproductive and respiratory syndrome virus. *Veterinary Research* 31:27-39.
- Gao, J., Yin, J., Xu, K., Li, T. and Yin, Y. (2019). 'What Is the Impact of Diet on Nutritional Diarrhea Associated with Gut Microbiota in Weaning Piglets: A System Review', *Hindawi BioMed Research International* Article ID 6916189, 1-14.
- Moeser, A. J., Pohl, C. S. and Rajput, M. (2017). 'Weaning stress and gastrointestinal barrier development: Implications for lifelong gut health in pigs', *Animal Nutrition* 3:313-321.
- Rhouma, M., Fairbrother, J. M., Beaudry, F. and Letellier, A. (2017). 'Post weaning diarrhea in pigs: risk factors and non-colistin-based control strategies', *Acta Veterinaria Scandinavica* 59:31.
- Rieckmann, K., Pendzialek, S. M., Vahlenkamp, T. and Baums, C. G. (2020). A critical review speculating on the protective efficacies of autogenous *Streptococcus suis* bacterins as used in Europe, *Porcine Health Management*, 6:12.

8.2 Key conferences held regularly every year or few years

- Allen D. Leman Swine Conference. https://ccaps.umn.edu/allen-d-leman-swine-conference.
- American Association of Swine Veterinarians. https://www.aasv.org.
- European Symposium of Porcine Health Management (ESPHM). https://ecphm.org/members-residents-area/meetings-trainings.
- International pig veterinary society (IPVS). https://www.theipvs.com.

8.3 Organisations and websites worth visiting where can be found research in this subject

- EU PiG Innovation Group (EU PiG). EU PiG aims are to help pig producers find tried-and-tested best practice from fellow producers across Europe, sharing all new knowledge in one place, online - funded by the European Horizon 2020 programme. https://www.eupig.co.uk.
- National Agricultural Library – U.S. Department of Agriculture. https://www.nal.usda.gov/topics/swine.
- Pig333. This website contains articles by swine specialists on health, management, environment, biosecurity, welfare, genetics, facilities and markets, as well as news, pig prices, clinical cases, events. https://www.pig333.com.
- Prohealth. The Prohealth Consortium has expertise in veterinary science and epidemiology, physiology and immunology, genetics, nutrition, socio-economics, welfare and production science of pigs and poultry https://www.fp7-prohealth.eu.
- The pig site. An updated guide focusing main subjects on pig health and management. https://www.thepigsite.com.

9 References

Alarcón, L. V., Allepuz, A. and Mateu, E. (2021). Biosecurity in pig farms: a review, *Porcine Health Manag.* 7(1):5. https://doi.org/10.1186/s40813-020-00181-z.

Albina, E., Madec, F., Cariolet, R. and Torrison, J. (1994). Immune response and persistence of the reproductive and respiratory syndrome virus in infected pigs and farm units, *Vet. Rec.* 134(22):567–573.

Albina, E., Mesplède, A., Chenut, G., Le Potier, M. F., Bourbao, G., Le Gal, S. and Leforban, Y. (2000). A serological survey on classical swine fever (CSF), Aujeszky's disease (AD) and porcine reproductive and respiratory syndrome (PRRS) virus infections in French wild boars from 1991 to 1998, *Vet. Microbiol.* 77(1–2):43–57.

Alexandersen, S., Knowles, N. J., Belsham, G. J., Dekker, A., Nfon, C., Zhang, Z. and Koenen, F. (2019). Picornaviruses. In: Zimmerman, J. J., Karriker, L. A., Ramirez, A., Schwartz, K. J., Stevenson, G. W. and Zhang, J. (Eds), *Disease of Swine* (11th edn.), Wiley-Blackwell, Hoboken, NJ, pp. 641–684.

Allwin, B., Swaminathan, R., Mohanraj, A., Suhas, G. N., Vedaminckam, S., Gopal, S. and Kumar, M. (2016). The wild pig (Sus scrofa) behavior - a retrospective study, *J. Vet. Sci. Techno* 7(4):333.

Almeida, F. N. and Stein, H. H. (2010). Performance and phosphorus balance of pigs fed diets formulated on the basis of values for standardized total tract digestibility of phosphorus, *J. Anim. Sci.* 88(9):2968-2977.

Alsaffar, A. A. (2010). Effect of thermal processing and storage on digestibility of starch in whole wheat grains, *J. Cereal Sci.* 52(3):480-485.

Aragon, V., Segalés, J. and Oliveira, S. (2012). Glässer disease. In: Zimmerm, J., Karrker, L., Ramirez, A., Schwarz, K. and Stevenson, G. (Eds), *Diseases of Swine* (10th edn.), Wiley-Blackwell, Ames, IA; The Atrium, Southern Gate, Chichester: Oxford, pp. 760-770.

Aragon, V., Segalés, J. and Tucker, A. W. D. (2019). Glässer disease. In: Zimmerman, J. J., Karriker, L. A., Ramirez, A., Schwartz, K. J., Stevenson, G. W. and Zhang, J. (Eds), *Disease of Swine* (11th edn.), Wiley-Blackwell, Hoboken, NJ, pp. 844-853.

Averos, X., Brossard, L., Dourmad, J. Y., de Greef, K. H., Edwards, S. A. and Meunier-Salaun, M. C. (2012). Meta-analysis on the effects of the physical environment, animal traits, feeder and feed characteristics on the feeding behaviour and performance of growing finishing pigs, *Animal* 6(8):1275-1289.

Bailey, M., Plunkett, F. J., Rothkotter, H. J., Vega-Lopez, M. A., Haverson, K. and Stokes, C. R. (2001). Regulation of mucosal immune responses in effector sites, *Proc. Nutr. Soc.* 60(4):427-435.

Bailey, M., Haverson, K., Inman, C., Harris, C., Jones, P., Corfield, G., Miller, B. and Stokes, C. (2005). The influence of environment on development of the mucosal immune system, *Vet. Immunol. Immunopathol.* 108(1-2):189-198.

Barcelo, J. and Marco, E. (1988). On-farm biosecurity. In: Varley, M. A., Done, S., Thomson, J. and International Pig Veterinary Society (Eds), *Scientific Committee of the 15th IPVS Congress. Proceedings of the 15th IPVS Congress*, Nottingham University Press. Nottingham (United Kingdom), pp. 129-133.

Baroch, J. A., Gagnon, C. A., Lacouture, S. and Gottschalk, M. (2015). Exposure of feral swine (Sus scrofa) in the United States to selected pathogens, *Can. J. Vet. Res.* 79(1):74-78.

Baums, C. G., Brüggemann, C., Kock, C., Beineke, A., Waldmann, K. H. and Valentin-Weigand, P. (2010). Immunogenicity of an autogenous *Streptococcus suis* bacterin in preparturient sows and their piglets in relation to protection after weaning, *Clin. Vaccine Immunol.* 17(10):1589-1597.

Bitsouni, V., Lycett, S., Opriessnig, T. and Doeschl-Wilson, A. (2019). Predicting vaccine effectiveness in livestock populations: a theoretical framework applied to PRRS virus infections in pigs, *PLoS ONE* 14(8):e0220738. https://doi.org/ 10.1371/journal.pone .0220738.

Boddicker, N., Waide, E. H., Rowland, R. R. R., Lunney, J. K., Garrick, D. J., Reecy, J. M. and Dekkers, J. C. M. (2012). Evidence for a major QTL associated with host response to porcine reproductive and respiratory syndrome virus challenge, *J. Anim. Sci.* 90(6):1733-1746.

Bonilauri, P., Merialdi, G., Dottori, M. and Barbieri, I. (2006). Presence of PRRSV in wild boar in Italy, *Vet. Rec.* 158(3):107-108.

Bottoms, K., Poljak, Z., Dewey, C., Deardon, R., Holtkamp, D. and Friendship, R. (2013). Evaluation of external biosecurity practices on Southern Ontario sow farms, *Prev. Vet. Med.* 109(1-2):58-68.

Boudry, G., Péron, V., Le Huérou-Luron, I., Lallès, J. P. and Sève, B. (2004). Weaning induces both transient and long-lasting modifications of absorptive, secretory, and barrier properties of piglet intestine, *J. Nutr.* 134(9):2256–2262.

Brean, M., Abraham, S., Hebart, M. and Kirkwood, R. N. (2016). Influence of parity of birth and suckled sows on piglet nasal mucosal colonization with *Haemophilus parasuis*, *Can. Vet. J.* 57(12):1281–1283.

Brockmeier, S. L., Register, K. B., Nicholson, T. L. and Loving, C. L. (2019). Bordetellosis. In: Zimmerman, J. J., Karriker, L. A., Ramirez, A., Schwartz, K. J., Stevenson, G. W. and Zhang, J. (Eds), *Disease of Swine* (11th edn.), Wiley-Blackwell, Hoboken, NJ, pp. 767–777.

Brown, L. G., Hoover, E. R., Selman, C. A., Coleman, E. W. and Schurz Rogers, H. (2017). Outbreak characteristics associated with identification of contributing factors to foodborne illness outbreaks, *Epidemiol. Infect.* 145(11):2254–2262.

Bruininx, E. M., Binnendijk, G. P., van der Peet-Schwering, C. M., Schrama, J. W., den Hartog, L. A., Everts, H. and Beynen, A. C. (2002). Effect of creep feed consumption on individual feed intake characteristics and performance of group-housed weanling pigs, *J. Anim. Sci.* 80(6):1413–1418.

Brumm, M. C. (2019). Effect of environment on health. In: Zimmerman, J. J., Karriker, L. A., Ramirez, A., Schwartz, K. J., Stevenson, G. W. and Zhang, J. (Eds), *Disease of Swine* (11th edn.), Wiley-Blackwell, Hoboken, NJ, pp. 50–58.

Cabrera, R. A., Usry, J. L., Arrellano, C., Nogueira, E. T., Kutschenko, M., Moeser, A. J. and Odle, J. (2013). Effect of creep feeding and supplemental glutamine or glutamine plus glutamate (Amino gut) on pre- and post-weaning growth performance and intestinal health of piglets, *J. Anim. Sci. Biotechnol.* 4(1):29.

Calsamiglia, M., Fraile, L., Espinal, A., Cuxart, A., Seminati, C., Martín, M., Mateu, E., Domingo, M. and Segalés, J. (2007). Sow porcine circovirus type 2 (PCV2) status effect on litter mortality in postweaning multisystemic wasting syndrome (PMWS), *Res. Vet. Sci.* 82(3):299–304.

Campbell, J. M., Crenshaw, J. D. and Polo, J. (2013). The biological stress of early weaned piglets, *J. Anim. Sci. Biotechnol.* 4(1):19.

Castillo, M., Martin-Orue, S. M., Nofrarias, M., Manzanilla, E. G. and Gasa, J. (2007). Changes in caecal microbiota and mucosal morphology of weaned pigs, *Vet. Microbiol.* 124(3–4):239–247.

Cerdà-Cuéllar, M., Naranjo, J. F., Verge, A., Nofrarias, M., Cortey, M., Olvera, A., Segales, J. and Aragon, V. (2010). Sow vaccination modulates the colonization of piglets by *Haemophilus parasuis*, *Vet. Microbiol.* 145(3–4):315–320.

Cevallos-Almeida, M., Fablet, C., Houdayer, C., Dorenlor, V., Eono, F., Denis, M. and Kerouanton, A. (2019). Longitudinal study describing time to Salmonella seroconversion in piglets on three farrow-to-finish farms, *Vet. Rec. Open* 6(1):e000287. https://doi.org/10.1136/vetreco-2018-000287.

Chae, C. (2012). Commercial porcine circovirus type 2 vaccines: efficacy and clinical application, *Vet. J.* 194(2):151–157.

Chae, C. (2021). Commercial PRRS modified-live virus vaccines, *Vaccines* 9(2):185. https://doi.org/10.3390/vaccines9020185.

Charerntantanakul, W. (2012). Porcine reproductive and respiratory syndrome virus vaccines: immunogenicity, efficacy and safety aspects, *World J. Virol.* 1(1):23–30. https://doi.org/10.5501/wjv.v1.i1.23.

Chase, C. and Lunney, J. K. (2019). Immune system. In: Zimmerman, J. J., Karriker, L. A., Ramirez, A., Schwartz, K. J., Stevenson, G. W. and Zhang, J. (Eds), *Disease of Swine* (11th edn.), Wiley-Blackwell, Hoboken, NJ, pp. 264-291.

Chen, H., Mao, X., He, J., Yu, B., Huang, Z., Yu, J., Zheng, P. and Chen, D. (2013). Dietary fibre affects intestinal mucosal barrier function and regulates intestinal bacteria in weaning piglets, *Br. J. Nutr.* 110(10):1837-1848.

Christianson, W. T., Choi, C. S., Collins, J. E., Molitor, T. W., Morrison, R. B. and Joo, H. S. (1993). Pathogenesis of porcine reproductive and respiratory syndrome virus infection in mid-gestation sows and fetuses, *Can. J. Vet. Res.* 57(4):262-268.

Cloutier, G., D'Allaire, S., Martinez, G., Surprenant, C., Lacouture, S. and Gottschalk, M. (2003). Epidemiology of *Streptococcus suis* serotype 5 infection in a pig herd with and without clinical disease, *Vet. Microbiol.* 97(1-2):135-151. https://doi.org/10.1016/j.vetmic.2003.09.018. PMID: 14637045.

Collins, C. L., Pluske, J. R., Morrison, R. S., McDonald, T. N., Smits, R. J., Henman, D. J., Stensland, I. and Dunshea, F. R. (2017). Post-weaning and whole-of-life performance of pigs is determined by live weight at weaning and the complexity of the diet fed after weaning, *Anim. Nutr.* 3(4):372-379.

Correa-Fiz, F., Fraile, L. and Aragon, V. (2016). Piglet nasal microbiota at weaning may influence the development of Glasser's disease during the rearing period, *BMC Genomics* 17:404.

Correa-Fiz, F., Neila-Ibáñez, C., López-Soria, S., Napp, S., Martinez, B., Sobrevia, L., Tibble, S., Aragon, V. and MiguraGarcia, L. (2020). Feed additives for the control of post-weaning *Streptococcus suis* disease and the effect on the faecal and nasal microbiota, *Sci. Rep.* 10(1):20354. https://doi.org/10.1038/s41598-020-77313-6.

Costa-Hurtado, M. and Aragon, V. (2013). Advances in the quest for virulence factors of *Haemophilus parasuis*, *Vet. J.* 198(3):571-576.

Costa-Hurtado, M., Barba-Vidal, E., Maldonado, J. and Aragon, V. (2020). Update on Glässer's disease: how to control the disease under restrictive use of antimicrobials, *Vet. Microbiol.* 242:108595. ISSN 0378-1135. https://doi.org/10.1016/j.vetmic.2020.108595.

Coutellier, L., Arnould, C., Boissy, A., Orgeur, P., Prunier, A., Veissier, I. and Meunier-Salaün, M. (2007). Pig's responses to repeated social regrouping and relocation during the growing-finishing period, *Appl. Anim. Behav. Sci.* 105(1-3):102-114.

Dee, S. A. and Corey, M. M. (1993). The survival of *Streptococcus suis* on farm and veterinary equipment, *Swine Health Prod.* 1(1):17-20.

Dee, S. A., Joo, H. S. and Pijoan, C. (1994). Controlling the spread of PRRS virus in the breeding herd through management of the gilt pool, *Swine Health Prod.* 3(2):64-69.

Dee, S. A., Joo, H. S., Polson, D. D., Park, B. K., Pijoan, C., Molitor, T. W., Collins, J. E. and King, V. (1997). Evaluation of the effects of nursery depopulation on the persistence of porcine reproductive and respiratory syndrome virus and the productivity of 34 farms, *Vet. Rec.* 140(10):247-248.

Dee, S. A., Deen, J., Burns, D., Douthit, G. and Pijoan, C. (2004). An assessment of sanitation protocols for commercial transport vehicles contaminated with porcine reproductive and respiratory syndrome virus, *Can. J. Vet. Res.* 68(3):208-214.

Dee, S. A., Torremorell, M., Thompson, B., Deen, J. and Pijoan, C. (2005). An evaluation of thermo-assisted drying and decontamination for the elimination of porcine reproductive and respiratory syndrome virus from contaminated livestock transport vehicles, *Can. J. Vet. Res.* 69(1):58-63.

Dee, S., Otake, S., Oliveira, S. and Deen, J. (2009). Evidence of long distance airborne transport of porcine reproductive and respiratory syndrome virus and *Mycoplasma hyopneumoniae*, *Vet. Res.* 40(4):39.

Dee, S. A., Otake, S. and Deen, J. (2011). An evaluation of ultraviolet light (UV 254) as a means to inactivate porcine reproductive and respiratory syndrome virus on common farm surfaces and materials, *Vet. Microbiol.* 150(1-2):96-99.

Devillers, N., Le Dividich, J. and Prunier, A. (2011). Influence of colostrum intake on piglet survival and immunity, *Animal* 5(10):1605-1612.

Done, S. H. and Paton, D. J. (1995). Porcine reproductive and respiratory syndrome: clinical disease, pathology and immunosuppression, *Vet. Rec.* 136(2):32-35.

Dong, G. Z. and Pluske, J. R. (2007). The low feed intake in newly-weaned pigs: problems and possible solutions, *Asian. Australas J. Anim. Sci.* 20(3):440-452.

Drew, T. W. (2000). A review of evidence for immunosuppression due to porcine reproductive and respiratory syndrome virus, *Vet. Res.* 31(1):27-39.

Dybkjær, L., Jacobsen, A. P., Tøgersen, F. A. and Poulsen, H. D. (2006). Eating and drinking activity of newly weaned piglets: effects of individual characteristics, social mixing, and addition of extra zinc to the feed, *J. Anim. Sci.* 84(3):702-711.

Enright, M. R., Alexander, T. J. L. and Clifton Hadley, F. A. (1987). Role of houseflies (*Musca domestica*) in the epidemiology of *Streptococcus suis* type 2, *Vet. Rec.* 121(6):132-133.

Fairbrother, J. M. and Nadeau, É. (2019). Colibacillosis. In: Zimmerman, J. J., Karriker, L. A., Ramirez, A., Schwartz, K. J., Stevenson, G. W. and Zhang, J. (Eds), *Disease of Swine* (11th edn.), Wiley-Blackwell, Hoboken, NJ, pp. 807-834.

FAO (Food and Agriculture Organization) of the United Nations/World Organisation for Animal Health/World Bank (2010). Good practices for biosecurity in the pig sector – issues and options in developing and transition countries. FAO Animal Production and Health Paper No. 169. Rome, FAO.

Feng, Y., Wang, Y., Wang, P., Huang, Y. and Wang, F. (2018). Short-chain fatty acids manifest stimulative and protective effects on intestinal barrier function through the inhibition of NLRP3 inflammasom and autophagy, *Cell. Physiol. Biochem.* 49(1):190-205.

Fleck, R. and Behrens, A. (2002). Evaluation of a maternal antibody decay curve for H1N1 swine influenza virus using the hemagglutination inhibition and the IDEXX ELISA tests. *Proceedings of the American Association of Swine Veterinarians* 2002: 109-110.

Fontanesi, L., Bertolini, F., Dall'Olio, S., Buttazzoni, L., Gallo, M. and Russo, V. (2012). Analysis of association between the MUC4 g.8227C>G polymorphism and production traits in Italian heavy pigs using a selective genotyping approach, *Anim. Biotechnol.* 23(3):147-155.

Franklin, M. A., Mathew, A. G., Vickers, J. R. and Clift, R. A. (2002). Characterization of microbial populations and volatile fatty acid concentrations in the jejunum, ileum, and cecum of pigs weaned at 17 vs 24 days of age, *J. Anim. Sci.* 80(11):2904-2910.

Gao, J., Yin, J., Xu, K., Li, T. and Yin, Y. (2019). What is the impact of diet on nutritional diarrhea associated with gut microbiota in weaning piglets: a system review, *Biomed Res. Int.* 2019:6916189

Gerritsen, R., van der Aar, P. and Molist, F. (2012). Insoluble nonstarch polysaccharides in diets for weaned piglets, *J. Anim. Sci.* 90 (Suppl. 4):318-320.

Gibson, G. R., Probert, H. M., Loo, J. V., Rastall, R. A. and Roberfroid, M. B. (2004). Dietary modulation of the human colonic microbiota: updating the concept of prebiotics, *Nutr. Res. Rev.* 17(2):259-275.

Go, N., Touzeau, S., Islam, Z., Belloc, C. and Doeschl-Wilson, A. (2019). How to prevent viremia rebound? Evidence from a PRRSv data-supported model of immune response, *BMC Syst. Biol.* 13(1):15. https://doi.org/10.1186/s12918-018-0666-7.

Gottschalk, M. and Broes, A. (2019). Actinobacillosis. In: Zimmerman, J. J., Karriker, L. A., Ramirez, A., Schwartz, K. J., Stevenson, G. W. and Zhang, J. (Eds), *Disease of Swine* (11th edn.), Wiley-Blackwell, Hoboken, NJ, pp. 749-766.

Gottschalk, M. and Segura, M. (2019). Streptococcosis. In: Zimmerman, J. J., Karriker, L. A., Ramirez, A., Schwartz, K. J., Stevenson, G. W. and Zhang, J. (Eds), *Disease of Swine* (11th edn.), Wiley-Blackwell, Hoboken, NJ, pp. 934-950.

Guilloteau, P., Martin, L., Eeckhaut, V., Ducatelle, R., Zabielski, R. and Van Immerseel, F. (2010). From the gut to the peripheral tissues: the multiple effects of butyrate, *Nutr. Res. Rev.* 23(2):366-384.

Guzman-Pino, S. A., Sola-Oriol, D., Davin, R., Manzanilla, E. G. and Perez, J. F. (2015). Influence of dietary electrolyte balance on feed preference and growth performance of post-weaned piglets, *J. Anim. Sci.* 93(6):2840-2848.

Hampson, D. J. and Kidder, D. E. (1986). Influence of creep feeding and weaning on brush border enzyme activities in the piglet small intestine, *Res. Vet. Sci.* 40(1):24-31.

Hancock, R. E. (2001). Cationic peptides: effectors in innate immunity and novel antimicrobials, *Lancet Infect. Dis.* 1(3):156-164.

He, Y., Deen, J., Shurson, G. C., Wang, L., Chen, C., Keisler, D. H. and Li, Y. Z. (2016). Identifying factors contributing to slow growth in pigs, *J. Anim. Sci.* 94(5):2103-2116.

Heo, J. M., Opapeju, F. O., Pluske, J. R., Kim, J. C., Hampson, D. J. and Nyachoti, C. M. (2013). Gastrointestinal health and function in weaned pigs: a review of feeding strategies to control post-weaning diarrhoea without using in-feed antimicrobial compounds, *J. Anim. Physiol. Anim. Nutr. (Berl)* 97(2):207-237.

Hou, Z., Wang, H., Feng, Y., Li, Q. and Li, J. (2019). A candidate DNA vaccine encoding a fusion protein of porcine complement C3d-P28 and ORF2 of porcine circovirus type 2 induces cross-protective immunity against PCV2b and PCV2d in pigs, *Virol. J.* 16:57.

Hyun, Y., Ellis, M., Riskowski, G. and Johnson, R. W. (1998). Growth performance of pigs subjected to multiple concurrent environmental stressors, *J. Anim. Sci.* 76(3):721-727.

Islam, Z. U., Bishop, S. C., Savill, N. J., Rowland, R. R., Lunney, J. K., Trible, B. and Doeschl-Wilson, A. B. (2013). Quantitative analysis of porcine reproductive and respiratory syndrome (PRRS) viremia profiles from experimental infection: a statistical modelling approach, *PLoS ONE* 8(12):e83567.

Jayaraman, B., Htoo, J. K. and Nyachoti, C. M. (2015). Effects of dietary threonine:lysine ratioes and sanitary conditions on performance, plasma urea nitrogen, plasma-free threonine and lysine of weaned pigs, *Anim Nutr.* 1(4):283-288.

Jayaraman, B., Htoo, J. K. and Nyachoti, C. M. (2017). Effects of different dietary tryptophan: lysine ratios and sanitary conditions on growth performance, plasma urea nitrogen, serum haptoglobin and ileal histomorphology of weaned pigs, *Anim. Sci. J.* 88(5):763-771.

Jayaraman, B. and Nyachoti, C. M. (2017). Husbandry practices and gut health outcomes in weaned piglets: a review, *Anim. Nutr.* 3(3):205-211.

Jeaurond, E. A., Rademacher, M., Pluske, J. R., Zhu, C. H. and de Lange, C. F. M. (2008). Impact of feeding fermentable proteins and carbohydrates on growth performance, gut health and gastrointestinal function of newly weaned pigs, *Can. J. Anim. Sci.* 88(2):271–281.

Jones, A. M., Wu, F., Woodworth, J. C., Dritz, S. S., Tokach, M. D., DeRouchey, J. M. and Goodband, R. D. (2019). Evaluation of dietary electrolyte balance on nursery pig performance, *Transl. Anim. Sci.* 3(1):378–383.

Jørgensen, C. B., Cirera, S., Anderson, S. I., Archibald, A. L., Raudsepp, T., Chowdhary, B., Edfors-Lilja, I., Andersson, L. and Fredholm, M. (2003). Linkage and comparative mapping of the locus controlling susceptibility towards E. coli F4ab/ac diarrhoea in pigs, *Cytogenet. Genome Res.* 102(1–4):157–162.

Karuppannan, A. K. and Opriessnig, T. (2018). *Lawsonia intracellularis*: revisiting the disease ecology and control of this fastidious pathogen in pigs, *Front. Vet. Sci.* 5:181.

Khafipour, E., Munyaka, P. M., Nyachoti, C. M., Krause, D. O. and Rodriguez-Lecompte, J. C. (2014). Effect of crowding stress and Escherichia coli K88+ challenge in nursery pigs supplemented with anti-Escherichia coli K88+ probiotics, *J. Anim. Sci.* 92(5):2017–2029.

Kirkland, P. D., Le Potier, M. F. and Finlaison, D. (2019). Pestiviruses. In: Zimmerman, J. J., Karriker, L. A., Ramirez, A., Schwartz, K. J., Stevenson, G. W. and Zhang, J. (Eds), *Disease of Swine* (11th edn.), Wiley-Blackwell, Hoboken, NJ, pp. 622–640.

Klinge, K. L., Vaughn, E. M., Roof, M. B., Bautista, E. M. and Murtaugh, M. P. (2009). Age-dependent resistance to porcine reproductive and respiratory syndrome virus replication in swine, *Virol. J.* 6:177.

Konstantinov, S. R., Awati, A. A., Williams, B. A., Miller, B. G., Jones, P., Stokes, C. R., Akkermans, A. D., Smidt, H. and de Vos, W. M. (2006). Post-natal development of the porcine microbiota composition and activities, *Environ. Microbiol.* 8(7):1191–1199.

Laine, T. M., Lyytikäinen, T., Yliaho, M. and Anttila, M. (2008). Risk factors for post-weaning diarrhoea on piglet producing farms in Finland, *Acta Vet. Scand.* 50(21):21.

Lallès, J. P., Bosi, P., Smidt, H. and Stoke, C. R. (2007). Nutritional management of gut health in pigs around weaning, *Proc. Nutr. Soc.* 66(2):260–268.

Langel, S. N., Paim, F. C., Lager, K. M., Vlasova, A. N. and Saif, L. J. (2016). Lactogenic immunity and vaccines for porcine epidemic diarrhea virus (PEDV): historical and current concepts, *Virus Res.* 226:93–107.

Lapointe, L., D'Allaire, S., Lebrun, A., Lacouture, S. and Gottschalk, M. (2002). Antibody response to an autogenous vaccine and serologic profile for *Streptococcus suis* capsular type 1/2, *Can. J. Vet. Res.* 66(1):8–14.

Le Floc'h, N., Lebellago, L., Matte, J.J., Melchior, D. and Seve, B. (2009). The effect of sanitary status degradation and dietary tryptophan content on growth rate and tryptophan metabolism in weaning pigs, *J. Anim. Sci.* 87:1686–1694.

Levis, D. G. and Baker, R. B. (2011). Biosecurity of pigs and farm security. *University of Nebraska Lincoln Extension Publications*. Available at: http://extesnion.unl.edu/publications.

Li, S., Sauer, W. C. and Fan, M. Z. (1993). The effect of dietary crude protein level on ileal and faecal amino acid digestibility in early-weaned pigs, *J. Anim. Physiol. Nutr.* 70(1–5):117–128.

Li, D. F., Jiang, J. Y. and Ma, Y. X. (2003). Early weaning diets and feed additives'. In: Xu, R. J. and Cranwell, P. D. (Eds). *The Neonatal Pig: Gastrointestinal Physiology and Nutrition*, Thrumptom, Nottingham. Nottingham University Press, p. 247.

Lindberg, J. E. (2014). Fiber effects in nutrition and gut health in pigs, *J. Anim. Sci. Biotechnol.* 5(1):15.

Liu, L., Wang, J., Zhao, Q., Zi, C., Wu, Z., Su, X., Huo, Y., Zhu, G., Wu, S. and Bao, W. (2013). Genetic variation in exon 10 of the BPI gene is associated with Escherichia coli F18 susceptibility in Sutai piglets, *Gene* 523(1):70-75.

Liu, Y., Espinosa, C. D., Abelilla, J. J., Casas, G. A., Lagos, L. V., Lee, S. A., Kwon, W. B., Mathai, J. K., Navarro, D. M. D. L., Hans, N. W. and Stein, H. H. (2018). Non-antibiotic feed additives in diets for pigs: a review, *Anim. Nutr.* 4(2):113-125.

López-Soria, S., Sibila, M., Nofrarías, M., Calsamiglia, M., Manzanilla, E. G., Ramírez-Mendoza, H., Mínguez, A., Serrano, J. M., Marín, O., Joisel, F., Charreyre, C. and Segalés, J. (2014). Effect of porcine circovirus type 2 (PCV2) load in serum on average daily weight gain during the postweaning period, *Vet. Microbiol.* 174(3-4):296-301.

Lordelo, M. M., Gaspar, A. M., Le Bellego, L. and Freire, J. P. B. (2008). Isoleucine and valine supplementation of a low-protein corn-wheat-soybean meal-based diet for piglets: growth performance and nitrogen balance, *J. Anim. Sci.* 86(11):2936-2941.

Luppi, A. (2017). Swine enteric colibacillosis: diagnosis, therapy and antimicrobial resistance, *Porcine Health Manag.* 3:16.

Maes, D., Verdonck, M., Deluyker, H. and de Kruif, A. (1996). Enzootic pneumonia in pigs, *Vet. Q.* 18(3):104-109.

Mahmmod, Y. S., Correa-Fiz, F. and Aragon, V. (2020). Variations in association of nasal microbiota with virulent and non-virulent strains of *Glaesserella* (*Haemophilus*) *parasuis* in weaning piglets, *Vet. Res.* 51(1):7. https://doi.org/10.1186/s13567-020-0738-8.

Marois, C., Le Devendec, L., Gottschalk, M. and Kobisch, M. (2007). Detection and molecular typing of *Streptococcus suis* in tonsils from live pigs in France, *Can. J. Vet. Res.* 71(1):14-22.

Martella, V., Banyai, K., Lorusso, E., Bellacicco, A. L., Decaro, N., Camero, M., Bozzo, G., Moschidou, P., Arista, S., Pezzotti, G., Lavazza, A. and Buonavoglia, C. (2007). Prevalence of group C rotaviruses in weaning and post-weaning pigs with enteritis, *Vet. Microbiol.* 123(1-3):26-33.

Martelli, P., Gozio, S., Ferrari, L., Rosina, S., De Angelis, E., Quintavalla, C., Bottarelli, E. and Borghetti, P. (2009). Efficacy of a modified live porcine reproductive and respiratory syndrome virus (PRRSV) vaccine in pigs naturally exposed to a heterologous European (Italian cluster) field strain: clinical protection and cell-mediated immunity, *Vaccine* 27(28):3788-3799. https://doi.org/10.1016/j.vaccine.2009.03.028 PMID: 19442420.

Martelli, P. (2013). Tabelle diagnosi differenziale. In: Martelli, P. (Ed.). *Le patologie del maiale*. Point Veterinaire Italie, Milano, pp. 2-5.

Martelli, P., Saleri, R., Ferrarini, G., De Angelis, E., Cavalli, V., Benetti, M., Ferrari, L., Canelli, E., Bonilauri, P., Arioli, E., Caleffi, A., Nathues, H. and Borghetti, P. (2016). Impact of maternally derived immunity on piglets' immune response and protection against porcine circovirus type 2 (PCV2) after vaccination against PCV2 at different age, *BMC Vet. Res.* 12:77.

Marthaler, D., Rossow, K., Culhane, M., Goyal, S., Collins, J., Matthijnssens, J., Nelson, M. and Ciarlet, M. (2014). Widespread rotavirus H in commercially raised pigs, United States, *Emerg. Infect. Dis.* 20(7):1195-1198.

Martinez, V., Wang, L., Million, M., Rivier, J. and Taché, Y. (2004). Urocortins and the regulation of gastrointestinal motor function and visceral pain, *Peptides* 25(10):1733-1744.

Mateos, G. G., López, E., Latorre, M. A., Vicente, B. and Lázaro, R. P. (2007). The effect of inclusion of oat hulls in piglet diet base on raw cooked rice and maize, *Anim. Feed Sci. Tech.* 135(1-2):100-112.

McGhee, J. R., Mestecky, J., Dertzbaugh, M. T., Eldridge, J. H., Hirasawa, M. and Kiyono, H. (1992). The mucosal immune system: from fundamental concepts to vaccine development, *Vaccine* 10(2):75–88.

McOrist, S., Bowles, R. and Blackall, P. (2009). Autogenous sow vaccination for Glasser's disease in weaner pigs in two large swine farm systems, *J. Swine Health Prod.* 17(2):90–96.

McRorie Jr., J. W. and McKeown, N. M. (2017). Understanding the physics of functional fibers in the gastrointestinal tract: an evidence-based approach to resolving enduring misconceptions about insoluble and soluble fiber, *J. Acad. Nutr. Diet.* 117(2):251–264.

Meerts, P., Misinzo, G., Lefebvre, D., Nielsen, J., Botner, A., Kristensen, C. S. and Nauwynck, H. J. (2006). Correlation between the presence of neutralizing antibodies against porcine circovirus 2 (PCV2) and protection against replication of the virus and development of PCV2-associated disease, *BMC Vet. Res.* 2:6.

Meijerink, E., Neuenschwander, S., Fries, R., Dinter, A., Bertschinger, H. U., Stranzinger, G. and Vögeli, P. (2000). A DNA polymorphism influencing (1,2) fucosyltransferase activity of the pig FUT1 enzyme determines susceptibility of small intestinal epithelium to Escherichia coli F18 adhesion, *Immunogenetics* 52(1–2):129–136.

Melkebeek, V., Goddeeris, B. M. and Cox, E. (2013). ETEC vaccination in pigs, *Vet. Immunol. Immunopathol.* 152(1–2):37–42.

Meng, X. J., Halbur, P. G. and Opriessnig, T. (2019). Hepatitis E. In: Zimmerman, J. J., Karriker, L. A., Ramirez, A., Schwartz, K. J., Stevenson, G. W. and Zhang, J. (Eds), *Disease of Swine* (11th edn.), Wiley-Blackwell, Hoboken, NJ, pp. 544–547.

Metzler, B. U., Vahjen, W., Baumgartel, T., Rodehutscord, M. and Mosenthin, R. (2009). Changes in bacterial populations in the ileum of pigs fed low-phosphorus diets supplemented with different sources of fermentable carbohydrates. *Anim. Feed Sci. Technol.* 148:69–89.

Moeser, A. J., Klok, C. V., Ryan, K. A., Wooten, J. G., Little, D., Cook, V. L. and Blikslager, A. T. (2007). Stress signaling pathways activated by weaning mediate intestinal dysfunction in the pig, *Am. J. Physiol. Gastrointest. Liver Physiol.* 292(1):G173–G181.

Moeser, A. J., Pohl, C. S. and Rajput, M. (2017). Weaning stress and gastrointestinal barrier development: implications for lifelong gut health in pigs, *Anim. Nutr.* 3(4):313–321.

Molist, F., de Segura, A. G., Gasa, J., Hermes, R. G., Manzanilla, E. G., Anguita, M. and Pérez, J. F. (2009). Effects of the insoluble and soluble dietary fibre on the physicochemical properties of digesta and the microbial activity in early weaned piglets, *Anim. Feed Sci. Technol.* 149(3–4):346–353.

Morris, C. R., Gardner, I. A., Hietala, S. K., Carpenter, T. E., Anderson, R. J. and Parker, K. M. (1994). Persistence of passively acquired antibodies to *Mycoplasma hyopneumoniae* in a swine herd, *Prev. Vet. Med.* 21(1):29–41.

Müller, T., Teuffert, J., Staubach, C., Selhorst, T. and Depner, K. R. (2005). Long-term studies on maternal immunity for Aujeszky's disease and classical swine fever in wild boar piglets, *J. Vet. Med. B Infect. Dis. Vet. Public Health* 52(10):432–436.

Murtaugh, M. P., Xiao, Z. and Zuckermann, F. (2002). Immunological responses of swine to porcine reproductive and respiratory syndrome virus infection, *Viral Immunol.* 15(4):533–547.

Nabuurs, M. J. A., Hoogendoorn, A., van der Molden, E. J. and Van Osta, A. L. M. (1993). Villous height and crypt depth in weaned and unweaned pigs, reared under various circumstances in the Netherlands, *Res. Vet. Sci.* 55(1):78–84.

Nagy, B., Nagy, G., Meder, M. and Mocsari, E. (1996). Enterotoxigenic Escherichia coli, rotavirus, porcine epidemic diarrhoea virus, adenovirus and calici-like virus in porcine postweaning diarrhoea in Hungary, *Acta Vet. Hung.* 44(1):9–19.

Noblet, J., Le Dividich, J. and Van Milgen, J. (2001). Thermal environment and swine nutrition. In: Lewis, A. J., Southern, L. L. and Swine Nutrition (Eds), 2nd ed., CRC Press LLC Press, Boca Raton, Palm Beach, FL, pp. 519–544.

NRC (2012). *Nutrient Requirements of Swine* (11th edn.), National Academic Press.

Nyachoti, C. M., Omogbenigun, F. O., Rademacher, M. and Blank, G. (2006). Performance responses and indicators of gastrointestinal health in early-weaned pigs fed low-protein amino acid-supplemented diets, *J. Anim. Sci.* 84(1):125–134.

Obradovic, M. R., Segura, M., Segalés, J. and Gottschalk, M. (2021). Review of the speculative role of co-infections in *Streptococcus suis*-associated diseases in pigs, *Vet. Res.* 52(1):49. https://doi.org/10.1186/s13567-021-00918-w.

O'Doherty, J. V., Bouwhuis, M. A. and Sweeney, T. (2017). Novel marine polysaccharides and maternal nutrition to stimulate gut health and performance in post-weaned pigs, *Anim. Prod. Sci.* 57(12):2376–2385. https://doi.org/10.1071/AN17272.

Oh, Y., Han, K., Seo, H. W., Park, C. and Chae, C. (2013). Program of vaccination and antibiotic treatment to control polyserositis caused by *Haemophilus parasuis* under field conditions, *Can. J. Vet. Res.* 77(3):183–190.

Ohland, C. L. and MacNaughton, W. K. (2010). Probiotic bacteria and intestinal epithelial barrier function, *Am. J. Physiol. Gastrointest. Liver Physiol.* 298(6):G807–G819.

Oliveira, S. and Pijoan, C. (2004). *Haemophilus parasuis*: new trends on diagnosis, epidemiology and control, *Vet. Microbiol.* 99(1):1–12.

Olvera, A., Sibila, M., Calsamiglia, M., Segalés, J. and Domingo, M. (2004). Comparison of porcine circovirus type 2 load in serum quantified by a real time PCR in postweaning multisystemic wasting syndrome and porcine dermatitis and nephropathy syndrome naturally affected pigs, *J. Virol. Methods* 117(1):75–80.

Opapeju, F. O., Krause, D. O., Payne, R. L., Rademacher, M. and Nyachoti, C. M. (2009). Effect of dietary protein level on growth performance, indicators of enteric health, and gastrointestinal microbial ecology of weaned pigs induced with postweaning colibacillosis, *Journal of Animal Science* 87:2635–2643.

Opapeju, F. O., Rademacher, M., Payne, R. L., Krause, D. O. and Nyachoti, C. M. (2010). Inflammation-associated responses in piglets induced with post-weaning colibacillosis are influenced by dietary protein level, *Livest. Sci.* 131(1):58–64.

Opriessnig, T., Jones, D., McKeown, N., Meng, X.-J. and Halbur, P. G. (2004). Development of a mouse model for PCV2-associated diseases. *Proceedings of the Con. Res. Work Animal Dis., Chicago, Illinois* 85:143.

Opriessnig, T., Giménez-Lirola, L. G. and Halbur, P. G. (2011). Polymicrobial respiratory disease in pigs, *Anim. Health Res. Rev.* 12(2):133–148.

Opriessnig, T., Gerber, P. F., Xiao, C. T., Halbur, P. G., Matzinger, S. R. and Meng, X. J. (2014). Commercial PCV2a-based vaccines are effective in protecting naturally PCV2b-infected finisher pigs against experimental challenge with a 2012 mutant PCV2, *Vaccine* 32(34):4342–4348.

Opriessnig, T. and Coutinho, T. A. (2019). Erysipelas. In: Zimmerman, J. J., Karriker, L. A., Ramirez, A., Schwartz, K. J., Stevenson, G. W. and Zhang, J. (Eds), *Disease of Swine* (11th edn.), Wiley-Blackwell, Hoboken, NJ, pp. 835–843.

Oslage, U., Dahle, J., Müller, T., Kramer, M., Beier, D. and Liess, B. (1994). Prevalence of antibodies against the viruses of European swine fever, Aujeszky's disease and porcine reproductive and respiratory syndrome in wild boars in the federal states Sachsen-Anhalt and Brandenburg, *Dtsch Tierarztl Wochenschr.* 101:33-38.

Ouyang, J., Zeng, W., Ren, J., Yan, X., Zhang, Z., Yang, M., Han, P., Huang, X., Ai, H. and Huang, L. (2012). Association of B3GNT5 polymorphisms with susceptibility to ETEC F4ab/ac in the white Duroc × Erhualian intercross and 15 outbred pig breeds, *Biochem. Genet.* 50(1-2):19-33.

Paul, P. S., Mengeling, W. L. and Pirtle, E. C. (1982). Duration and biological half-life of passively acquired colostral antibodies to porcine parvovirus, *Am. J. Vet. Res.* 43(8):1376-1379.

Pereira, D. A., Dalla Costa, F. A., Ferroni, L. B., Moraes, C. N., Schocken-Iturrino, R. P. and Oliveira, L. G. (2017). The challenges with Glässer's disease in technified pig production, *Aust. J. Vet. Sci.* 49:63-69.

Pileri, E. and Mateu, E. (2016). Review on the transmission porcine reproductive and respiratory syndrome virus between pigs and farms and impact on vaccination, *Vet. Res.* 47(1):108.

Pinilla, J. C., Geiger, J., Kummer, R., Piva, J., Schott, R. and Williams, N. H. (2008). *Management Strategies to Maximize Weaning Weight*, American Association Of Swine Veterinarians, pp. 185-192.

Pluske, J. R., Williams, I. H. and Aherne, F. X. (1996). Maintenance of villous height and crypt depth in piglets by providing continuous nutrition after weaning, *Anim. Sci.* 62(1):131-144.

Pluske, J. R., Hampson, D. J. and Williams, I. H. (1997). Factors influencing the structure and function of the small intestine in the weaned pig: a review, *Livest. Prod. Sci.* 51(1-3):215-236.

Pluske, J. R., Pethick, D. W., Hopwood, D. E. and Hampson, D. J. (2002). Nutritional influences on some major enteric bacterial diseases of pigs, *Nutr. Res. Rev.* 15(2):333-371.

Pomorska-Mol, M., Markowska-Daniel, I., Rachubik, J. and Pejsak, Z. (2011). Effect of maternal antibodies and pig age on the antibody response after vaccination against Glässer's disease, *Vet. Res. Commun.* 35(6):337-343.

Poonsuk, K. and Zimmerman, J. (2018). Historical and contemporary aspects of maternal immunity in swine, *Anim. Health Res. Rev.* 19(1):31-45.

Quesnel, H., Brossard, L., Valancogne, A. and Quiniou, N. (2008). Influence of some sow characteristics on within-litter variation of piglet birth weight, *Animal* 2(12):1842-1849. HTTPS://doi.org/10.1017/S175173110800308X.

Racine, N. M., Riddell, R. R., Khan, M., Calic, M., Taddio, A. and Tablon, P. (2016). Systematic review: predisposing, precipitating, perpetuating, and present factors predicting anticipatory distress to painful medical procedures in children, *J. Pediatr. Psychol.* 41(2):159-181.

Rattigan, R., Sweeney, T., Maher, S., Ryan, M. T., Thornton, K. and O'Doherty, J. V. (2020). Effects of reducing dietary crude protein concentration and supplementation with either laminarin or zinc oxide on the growth performance and intestinal health of newly weaned pigs, *Anim. Feed Sci. Technol.* 270:114693. https://doi.org/10.1016/j.anifeedsci.2020.114693.

Reiner, G., Fresen, C., Bronnert, S. and Willems, H. (2009). Porcine reproductive and respiratory syndrome virus (PRRSV) infection in wild boars, *Vet. Microbiol.* 136(3-4):250-258.

Rhouma, M., Fairbrother, J. M., Beaudry, F. and Letellier, A. (2017). Post weaning diarrhea in pigs: risk factors and non-colistin-based control strategie's, *Acta Vet. Scand.* 59(1):31.

Riaz Rajoka, M. S., Shi, J., Mehwish, H. M., Zhu, J., Li, Q., Shao, D., Huang, Q. and Yang, H. (2017). Interaction between diet composition and gut microbiota and its impact on gastrointestinal tract health, *Food Sci. Hum. Wellness* 6(3):121-130.

Rieckmann, K., Pendzialek, S. M., Vahlenkamp, T. and Baums, C. G. (2020). A critical review speculating on the protective efficacies of autogenous *Streptococcus suis* bacterins as used in Europe, *Porcine Health Manag.* 6:12.

Robinson, K. M. X., Liu, Y., Qiao, S., Hou, Y. and Zhang, G. (2018). Dietary modulaton of endogenois host defense peptides synthesis as an alternative approach to infeed antibiotics, *Anim. Nutr.* 4:160-169.

Roca, M., Gimeno, M., Bruguera, S., Segalés, J., Díaz, I., Galindo-Cardiel, I. J., Martínez, E., Darwich, L., Fang, Y., Maldonado, J., March, R. and Mateu, E. (2012). Effect of challenge with a virulent genotype II strain of porcine reproductive and respiratoiry syndrome virus on piglets vaccinated with an attenuated genotype I strain vaccine, *Vet. J.* 193(1):92-96.

Rodríguez-Arrioja, G. M., Segalés, J., Calsamiglia, M., Resendes, A. R., Balasch, M., Plana-Durán, J., Casal, J. and Domingo, M. (2002). Dynamics of porcine circovirus type 2 infection in a herd of pigs with postweaning multisystemic wasting syndrome, *Am. J. Vet. Res.* 63(3):354-357.

Roic, B., Jemersic, L., Terzic, S., Keros, T., Balatinec, J. and Florijancic, T. (2012). Prevalence of antibodies to selected viral pathogens in wild boars (Sus scrofa) in Croatia in 2005-2006 and 2009-2010, *J. Wild Dis.* 48:131-137.

Rose, N. and Andraud, M. (2017). The use of vaccines to control pathogen spread in pig populations, *Porcine Health Manag.* 3:8 https://doi.org/10.1186/s40813-017-0053-6.

Roselli, M., Finamore, A., Garaguso, I., Britti, M. S. and Mengheri, E. (2003). Zinc oxide protects cultured enterocytes from the damage induced by *Escherichia coli*, *J. Nutr.* 133(12):4077-4082.

Saif, L. J., Wang, Q., Vlasova, A. N., Jung, K. and Xiao, S. (2019). Coronaviruses. In: Zimmerman, J. J., Karriker, L. A., Ramirez, A., Schwartz, K. J., Stevenson, G. W. and Zhang, J. (Eds), *Disease of Swine* (11th edn.), Wiley-Blackwell, Hoboken, NJ, pp. 488-523.

Sakata, T., Kojima, T., Fujieda, M., Takahashi, M. and Michibata, T. (2003). Influences of probiotic bacteria on organic acid production by pig caecal bacteria in vitro, *Proc. Nutr. Soc.* 62(1):73-80.

Saravanan, S., Geurden, I., Orozco, Z. G., Kaushik, S. J., Verreth, J. A. and Schrama, J. W. (2013). Dietary electrolyte balance affects the nutrient digestibility and maintenance energy expenditure of Nile tilapia, *Br. J. Nutr.* 110(11):1948-1957.

Schurrer, J. A., Dee, S. A., Moon, R. D., Murtaugh, M. P., Finnegan, C. P., Deen, J., Kleiboeker, S. B. and Pijoan, C. B. (2005). Retention of ingested porcine reproductive and respiratory syndrome virus in houseflies, *Am. J. Vet. Res.* 66(9):1517-1525.

Segalés, J. and Domingo, M. (2002). Postweaning multisystemic wasting syndrome (PMWS) in pigs: a review, *Vet. Q.* 24(3):109-124.

Segalés, J., Domingo, M., Chianini, F., Majo, N., Dominguez, J., Darwich, L. and Mateu, E. (2004). Immunosuppression in postweaning multisystemic wasting syndrome affected pigs, *Vet. Microbiol.* 98(2):151-158.

Segalés, J. (2015). Best practice and future challenges for vaccination against porcine circovirus type 2', *Vaccines* 14(3):473-487.

Segalés, J. (2020). Porcine circoviruses in 2020: what's new? In: Stingelin, G. M., Guilherme de Oliveira, L. and Montenegro Franceschini, V. (Eds) *Swine Health and Production: Updating, Innovation and Technology*, FUNEP Jaboticabal, São Paulo, pp. 4-9.

Segura, M. (2015). *Streptococcus suis* vaccines: candidate antigens and progress, *Expert Rev. Vaccines* 14(1):1–22.

Segura, M., Calzas, C., Grenier, D. and Gottschalk, M. (2016). Initial steps of the pathogenesis of the infection caused by *Streptococcus suis*: fighting against nonspecific defenses, *FEBS Lett.* 590(21):3772–3799.

Segura, M., Aragon, V., Brockmeier, S. L., Gebhart, C., de Gree, A., Kerdsin, A., O'Dea, M. A., Okura, M., Saléry, M., Schultsz, C., Valentin-Weigand, P., Weinert, L. A., Wells, J. M. and Gottschalk, M. (2020). Update on *Streptococcus suis* research and prevention in the era of antimicrobial restriction: 4th international workshop on *S. suis*. Pathogens, 9(5):374.

Seo, H. W., Park, C., Han, K. and Chae, C. (2014). Effect of porcine circovirus type 2 (PCV2) vaccination on PCV2-viremic piglets after experimental PCV2 challenge, *Vet. Res.* 45:13.

Shen, H. G., Loiacono, C. M., Halbur, P. G. and Opriessnig, T. (2012). Age-dependent susceptibility to porcine circovirus type 2 infections is likely associated with declining levels of maternal antibodies, *J. Swine Health Prod.* 20(1):17–24.

Shepherd, F. K., Freeman, M. J., Culhane, M. R. and Marthaler, D. G. (2019). Reoviruses (rotaviruses and Reoviruses). In: Zimmerman, J. J., Karriker, L. A., Ramirez, A., Schwartz, K. J., Stevenson, G. W. and Zhang, J. (Eds), *Disease of Swine* (11th edn.), Wiley-Blackwell, Hoboken, NJ, pp. 715–727.

Shirai, J., Kanno, T., Tsuchiya, Y., Mitsubayashi, S. and Seki, R. (2000). Effects of chlorine, iodine, and quaternary ammonium compound disinfectants on several exotic disease viruses, *J. Vet. Med. Sci.* 62(1):85–92.

Silva, J., Rocha, D., Cunha, I., Rui Sales, L., Neto, F., Fontes, M. C. and Simões, J. (2015). Serological profile of offspring on an intensive pig farm affected by porcine reproductive and respiratory syndrome, *Asian Pac. J. Reprod.* 4(4):317–321.

Snijder, E. J. and Meulenberg, J. M. (2001). Arteriviruses. In: Knipe, D. M., Howley, P. M., Griffin, D. E., et al. (Eds), *Fields Virology* (4th edn.), Lippincott Williams and Wilkins, Philadelphia, PA, pp. 1205–1220.

Stein, H. H. and Kil, D. Y. (2006). Reduced use of antibiotic growth promoters in diets fed to weanling pigs: dietary tools, part 2, *Anim. Biotechnol.* 17(2):217–231.

Stokes, C. R. (2017). The development and role of microbial-host interactions in gut mucosal immune development, *J. Anim. Sci. Biotechnol.* 8:12.

Swildens, B., Nielen, M., Wisselink, H. J., Verheijden, J. H. and Stegeman, J. A. (2007). Elimination of strains of *Streptococcus suis* serotype 2 from the tonsils of carrier sows by combined medication and vaccination, *Vet. Rec.* 160(18):619–621.

Thomas, C., Mettenleiter, B. E., Müller, T., Yoon, K. J. and Teifke, J. P. (2019). Herpesviruses. In: Zimmerman, J. J., Karriker, L. A., Ramirez, A., Schwartz, K. J., Stevenson, G. W. and Zhang, J. (Eds), *Disease of Swine* (11th edn.), Wiley-Blackwell, Hoboken, NJ, pp. 548–575.

Truyen, U. and Streck, A. F. (2019). *Parvoviruses*. In: Zimmerman, J. J., Karriker, L. A., Ramirez, A., Schwartz, K. J., Stevenson, G. W. and Zhang, J. (Eds), *Disease of Swine* (11th edn.), Wiley-Blackwell, Hoboken, NJ, pp. 611–621.

Tsiloyiannis, V. K., Kyriakis, S. C., Vlemmas, J. and Sarris, K. (2001). The effect of organic acids on the control of porcine post-weaning diarrhoea, *Res. Vet. Sci.* 70(3):287–293.

Tzipori, S., Chandler, D., Makin, T. and Smith, M. (1980). *Escherichia coli* and Rotavirus infections in four-week-old gnotobiotic piglets fed milk or dry food, *Aust. Vet. J.* 56(6):279–284.

Van Reeth, K. and Vincent, A. L. (2019). Influenza virus. In: Zimmerman, J. J., Karriker, L. A., Ramirez, A., Schwartz, K. J., Stevenson, G. W. and Zhang, J. (Eds), *Disease of Swine* (11th edn.), Wiley-Blackwell, Hoboken, NJ, pp. 576–593.

Veldkamp, T. and Vernooij, A. G. (2021). Use of insect products in pig diets, *J. Insects Food Feed* 0(0):1–14. https://doi.org/10.3920/JIFF2020.0091. in Press.

Vente-Spreeuwenberg, M. A. M., Verdonk, J. M. A. J., Gaskins, H. J. and Verstegen, M. W. A. (2001). Small intestine epithelial barrier function is compromised in pigs with low feed intake at weaning, *J. Anim. Nutr.* 131:1520–1527.

Vente-Spreeuwenberg, M. A. M., Verdonk, J. M. A. J., Verstegen, M. W. A. and Beynen, A. C. (2003). Villus height and gut development in weaned piglets receiving diets containing either glucose, lactose or starch, *Br. J. Nutr.* 90(5):907–913.

Vente-Spreeuwenberg, M. A. M., Verdonk, J. M. A. J., Bakker, G. C. M., Beynen, A. C. and Verstegen, M. W. A. (2004). Effect of dietary protein source on feed intake and small intestinal morphology in newly weaned piglets, *Livest. Prod. Sci.* 86(1–3):169–177.

Vilcek, S., Molnar, L., Vlasakova, M. and Jackova, A. (2015). The first detection of PRRSV in wild boars in Slovakia, *Berl. Munch. Tierarztl Wochenschr.* 128(1–2):31–33.

Vötsch, D., Willenborg, M., Weldearegay, Y. B. and Valentin-Weigand., P. (2018). *Streptococcus suis* – the "two faces" of a pathobiont in the porcine respiratory tract, *Front. Microbiol.* 9:480.

Wang, Y., Shan, T., Xu, Z., Liu, J. and Feng, J. (2006). Effect of lactoferrin on the growth performance, intestinal morphology, and expression of PR-39 and protegrin-1 genes in weaned piglets, *J. Anim. Sci.* 84(10):2636–2641.

Wang, Y., Ren, J., Lan, L., Yan, X., Huang, X., Peng, Q., Tang, H., Zhang, B., Ji, H. and Huang, L. (2007). Characterization of polymorphisms of transferrin receptor and their association with susceptibility to ETEC F4ab/ac in pigs, *J. Anim. Breed. Genet.* 124(4):225–229.

Wang, J., Chen, L., Li, P., Li, X., Zhou, H., Wang, F., Li, D., Yin, Y. and Wu, G. (2008). Gene expression is altered in piglet small intestine by weaning and dietary glutamine supplementation, *J. Nutr.* 138(6):1025–1032.

Wang, J., Li, G. R., Tan, B. E., Xiong, X., Kong, X. F., Xiao, D. F., Xu, L. W., Wu, M. M., Huang, B., Kim, S. W. and Yin, Y. L. (2015). Oral administration of putrecine and proline during the suckling period improve epithelial restitution after early weaning in piglets, *J. Anim. Sci.* 93(4):1679–1688.

Warneboldt, F., Sander, S. J., Beineke, A., Valentin-Weigand, P., Kamphues, J. and Baums, C. G. (2016). Clearance of *Streptococcus suis* in stomach contents of differently fed growing pigs, *Pathogens* 5(3):56.

Wen, Z.-S., Lu, J.-L. and Zou, X.-T. (2012). Effect of sodium butyrate on the intestinal morphology and DNA-binding activity of intestinal nuclear factor-κB in weanling pigs, *J. Anim. Vet. Adv.* 11(6):814–821.

Williams, B. A., Verstegen, M. W. A. and Tamminga, S. (2001). Fermentation in the monogastric large intestine: its relation to animal health, *Nutr. Res. Rev.* 14(2):207–228.

Wolf, J., Žáková, E. and Groeneveld, E. (2008). Within-litter variation of birth weight in hyperprolific Czech Large White sows and its relation to litter size traits, stillborn piglets and losses until weaning, *Livest. Sci.* 115(2–3):195–205.

Wu, G., Fang, Y. Z., Yang, S., Lupton, J. R. and Turner, N. D. (2004). Glutathione metabolism and its implications for health, *J. Nutr.* 134(3):489–492.

Wu, Y. P., Jiang, Z. Y., Zheng, C. T., Wang, L., Zhu, C., Yang, X. F., Wen, X. L. and Ma, X. Y. (2015). Effects of protein sources and levels in antibiotic-free diets on diarrhea, intestinal morphology, and expression of tight junctions in weaned piglets, *Anim. Nutr.* 1(3):170-176.

Wu, M., Xiao, H., Liu, G., Chen, S., Tan, B., Ren, W., Bazer, F. W., Wu, G. and Yin, Y. (2016). Glutamine promotes intestinal SIgA secretion through intestinal microbiota and IL-13, *Mol. Nutr. Food Res.* 60(7):1637-1648.

Xiong, X., Tan, B., Song, M., Ji, P., Kim, K., Yin, Y. and Liu, Y. (2019). Nutritional intervention for the intestinal development and health of weaned pigs, *Front. Vet. Sci.* 6:46.

Yang, X. F., Jiang, Z. Y., Gong, Y. L., Zheng, C. T., Hu, Y. J., Wang, L., Huang, L. and Ma, X. Y. (2016). Supplementation of pre-weaning diet with L-arginine has carry-over effect to improve intestinal development in young piglets, *Can. J. Anim. Sci.* 96(1):52-59.

Yoon, K. J., Zimmerman, J. J., Swenson, S. L., McGinley, M. J., Eernisse, K. A., Brevik, A., Rhinehart, L. L., Frey, M. L., Hill, H. T. and Platt, K. B. (1995). Characterization of the humoral immune response to porcine reproductive and respiratory syndrome (PRRS) virus infection, *J. Vet. Diagn. Invest.* 7(3):305-312.

Zhang, B., Ren, J., Yan, X. M., Huang, X., Ji, H. Y., Peng, Q. L., Zhang, Z. Y. and Huang, L. S. (2008). Investigation of the porcine MUC13 gene: isolation, expression, polymorphisms and their strong association with susceptibility to 282 enterotoxigenic *E. coli* F4ab/ac, *Anim. Genet.* 39(3):258-266.

Zhao, Z., Liu, H., Xue, Y., Chen, K., Liu, Z., Xue, Q. and Wang, C. (2017). Analysis of efficacy obtained with a trivalent inactivated *Haemophilus parasuis* serovars 4, 5, and 12 vaccine and commercial vaccines against Glässer's disease in piglets, *Can. J. Vet. Res.* 81(1):22-27.

Zimmerman, J. J., Dee, S. A., Holtkamp, D. J., Murtaugh, M. P., Stadejek, T., Stevenson, G. W., Torremorell, M., Yang, H. and Zhang, J. (2019). Porcine reproductive and respiratory syndrome viruses (porcine arteriviruses). In: Zimmerman, J. J., Karriker, L. A., Ramirez, A., Schwartz, K. J., Zhang, J. and Stevenson, G. W. (Eds), *Disease of Swine* (11th edn.), Wiley-Blackwell, Hoboken, NJ, pp. 685-708.

Zong, E., Huang, P., Zhang, W., Li, J., Li, Y., Ding, X., Xiong, X., Yin, Y. and Yang, H. (2018). The effects of dietary sulfur amino acids on growth performance, intestinal morphology, enzyme activity, and nutrient transporters in weaning piglets, *J. Anim. Sci.* 96(3):1130-1139.

Zou, Y., Xiang, Q., Wang, J., Peng, J. and Wei, H. (2016). Oregano essential oil improves intestinal morphology and expression of tight junction proteins associated with modulation of selected intestinal bacteria and immune status in a pig model, *BioMed Res. Int.* 2016:5436738. https://doi.org/10.1155/2016/5436738.

Zulovich, J. M. (2012). Effect of the environment on health. In: Zimmerman, J., Karrker, L., Ramirez, A., Schwarz, K. and Stevenson, G. (Eds), *Diseases of Swine* (10th edn.), Wiley-Blackwell, Ames, IA; The Atrium, Southern Gate, Chichester; Oxford, pp. 60-66.

www.ingramcontent.com/pod-product-compliance
Lightning Source LLC
Chambersburg PA
CBHW050523270326
41926CB00015B/3046